BRAHAM TINDALE
41 WILCOX AVE
SINGLETON HEIGHTS
2330

Planning Occupational Health & Safety

5th Edition

Important Disclaimer

No person should rely on the contents of this publication without first obtaining advice from a qualified professional person. This publication is sold on the terms and understanding that (1) the authors, consultants and editors are not responsible for the results of any actions taken on the basis of information in this publication, nor for any error in or omission from this publication; and (2) the publisher is not engaged in rendering legal, accounting, professional or other advice or services. The publisher, and the authors, consultants and editors, expressly disclaim all and any liability and responsibility to any person, whether a purchaser or reader of this publication or not, in respect of anything, and of the consequences of anything, done or omitted to be done by any such person in reliance, whether wholly or partially, upon the whole or any part of the contents of this publication. Without limiting the generality of the above, no author, consultant or editor shall have any responsibility for any act or omission of any other author, consultant or editor.

4519A

Planning Occupational Health & Safety

5th Edition

CCH AUSTRALIA LIMITED
ACN 000 630 197
ABN 67 000 630 197
GPO Box 4072, Sydney, NSW 2001
Head Office North Ryde Phone: (02) 9857 1300 Fax: (02) 9857 1600
Customer Support Phone: 1 300 300 224 Fax: 1 300 306 224
www.cch.com.au

About CCH Australia Limited

CCH Australia Limited is part of a leading global organisation publishing in many countries.

CCH began business in Australia in 1969 and quickly established a solid reputation as the country's leading tax publisher. Today our scope is much broader, and our product range is constantly being expanded to meet customer needs.

CCH publications cover a wide variety of topical areas, including tax, company law, accounting, contract law, conveyancing, human resources, industrial law, occupational health and safety, torts and training. CCH is committed to identifying and integrating new technologies into its products.

Our clients include barristers, solicitors, accountants, human resources managers, OHS specialists, business people and students. We aim to provide up-to-date, accurate, authoritative, knowledge-based, practical information which customers can easily and quickly apply to their own specific circumstances.

Enquiries are welcome on 1300 300 224.

National Library of Australia Cataloguing in Publication Data

Planning occupational health and safety.
  5th ed.
  Includes index.

  ISBN 1 86468 455 0.

  1. Industrial safety – Australia. 2. Industrial hygiene – Australia. 3. Personnel Management – Australia – Planning. I. CCH Australia Limited.

363.110994

© 2001 CCH Australia Limited   ABN 67 000 630 197

First published ....February 1983	4th edition ....November 1996
2nd edition ....February 1987	5th edition ....June 2000
3rd edition ....December 1991	Reprinted with revisions ......April 2001

This publication was first entitled *Planning Occupational Safety & Health*

All rights reserved. No part of this work covered by copyright may be reproduced or copied in any form or by any means (graphic, electronic or mechanical, including photocopying, recording, recording taping, or information retrieval systems) without the written permission of the publisher. Wholly set up and printed in Australia.

Cover design and execution by Typezone, Lindfield.

Printed in Australia by Australian Print Group.

Foreword

Planning Occupational Health & Safety — 5th Edition is a practical guide for people employed in human resource departments, occupational health and safety professionals; students of the subject; and those starting up a new occupational health and safety (OHS) position or department in an organisation.

For a long time in Australia, a planned approach to workplace health and safety was largely seen as an "optional extra" — something to attend to after taking care of business. A reckoning of the costs flowing from this — to individual organisations and their employees, industry and society at large — has altered this perception. OHS is now an integral part of business operation, and planning here, as in other aspects of business, is recognised as essential.

This publication aims to describe how to introduce and maintain the OHS function in an organisation. The book includes discussion of the impact of legislation, factors to consider when planning and introducing the function, the role of various people in the organisation, administration and methods of control and review.

Chapter 1 examines the importance of planning OHS, including the cost of unsafe practices and the changes in approach occasioned by these costs. Chapter 2 provides an overview of the legislation dealing with health and safety in Australia.

Chapter 3 is an outline of the risk management process, which is the basis of the modern approach to OHS legislation. It provides practical advice on tackling each stage.

Chapters 4-8 discuss the infrastructure and planning needed to support the risk management process. This includes an assessment of the organisation's current position, staffing considerations, the role of safety representatives and committees, and techniques for implementing safety programs and investigating accidents at the workplace. Ongoing evaluation of OHS practice is also discussed.

Chapters 9-11 provide an introduction to examples of common health and safety problems, setting out some typical approaches to risk control in these areas.

Chapter 12 outlines further sources of information including government departments, private organisations and various publications.

For more detailed discussions on any of the topics raised in the book, the extensive range of CCH Australia's OHS publications are an essential guide.

Information on the state of the art in OHS practice (including information on conferences/seminars, videos, books and education) is published in CCH's bi-monthly magazine *The Journal of Occupational Health and Safety — Australia and New Zealand*.

CCH also publishes a three-volume reporting service, *Australian Occupational Health & Safety Law*, which provides the text of occupational health and safety legislation in each State and Territory. CCH's single-volume loose-leaf service *Managing Occupational Health & Safety*, is a practical on-site reference for those managing health and safety. For a more extensive list of CCH titles in the area of OHS see Chapter 12.

Acknowledgements

CCH Australia Limited acknowledges the contribution to the previous editions of this book by Richard M Ridout, FSIA, Leigh Deves BA (Hons), MBA, Glyn Williams and Joy Window, BSc (Hons), GradDipEditPub.

This edition was updated by Gaby Grammeno BA (Hons), MPH, with assistance from Dennis Hine Bbus, GradDip Ed (Tech) MEd (Adult Ed.), CMAHRI.

CCH Australia Limited

July 2000

Contents

	Page
Foreword	v
Chapter 1: Why plan for Occupational Health and Safety?	1
Chapter 2: Overview of Legislation	15
Chapter 3: Risk Management	43
Chapter 4: Planning a Health and Safety Program	61
Chapter 5: Staffing the Health and Safety Function	83
Chapter 6: Techniques of Accident Investigation, Prevention and Reporting	107
Chapter 7: Implementing a Program	123
Chapter 8: Evaluating Health and Safety Performance	151
Chapter 9: An Introduction to Health and Safety Problems — Work Environments	159
Chapter 10: An Introduction to Health and Safety Problems — Wellness in the Workplace	201
Chapter 11: An Introduction to Health and Safety Problems — Social Issues	239
Chapter 12: Reference Section	271
Index	284

Chapter 1

Why Plan for Occupational Health and Safety?

Introduction — various approaches to occupational health
and safety .... ¶101
Measuring the problem .... ¶102
Indirect costs .... ¶103
Changes .... ¶104
Management's role .... ¶105
Employee's role — developing a consultative approach ... ¶106
The role of the human resources department/health
and safety department .... ¶107
Aims of an occupational health and safety program .... ¶108
Some definitions .... ¶109
Scope of this book .... ¶110

¶101 Introduction — various approaches to occupational health and safety

Ask managers whether they are safety conscious and care about employee wellbeing and you will seldom receive a negative answer. In most cases, managers will genuinely believe this answer, but it may mean no more than:

- adequate workers compensation cover has been taken out;
- first aid facilities are provided;
- line managers are instructed to exercise sufficient care in discharging their duties to ensure management interests are protected;
- new equipment purchased is expected to meet the requirements of existing safety legislation; and
- anything of a serious nature is drawn to management's attention but the extent of action or investigation depends upon other priorities and

how important health and safety is considered to be in relation to those priorities.

This approach has been described as "minimum standards compliance". That is, the workplace meets minimum standards prescribed in the State's legislation in matters such as the fencing off of dangerous machinery. However, when assessed against the increasing costs of workplace injuries, this approach has been found wanting since:

- the causes of some common injuries are not amenable to solution by the old style of prescriptive regulation (for example, back injuries from slips and falls or from unsafe manual handling);
- many health problems are only now being identified as work-related, in whole or part, (for example, the effects of excessive exposure to sunlight); and
- there are obvious difficulties in prescribing standards to precisely meet the various conditions at different workplaces, and this old style of prescriptive legislation left many areas of the workforce unprotected by legislation.

> In contrast to the above approach, this book promotes an integrated approach to occupational health and safety. Health and safety become the concern of the whole organisation, with a particular emphasis on management's role of instigating and supporting health and safety programs. Prevention becomes the focus, rather than reaction to individual occurrences.

In Australia there has been a strong move towards a consultative approach to occupational health and safety. This means that management works with employees to create a safe workplace, with the final decision-making power resting with management. The consultative approach is supported by the modern style of occupational safety legislation, which contains provisions to establish health and safety representatives or committees. This legislation is detailed in Chapter 2. Strategies for introducing a consultative approach to occupational health and safety, as well as the advantages of such an approach, are discussed at ¶106.

The reasons for emphasising an integrated, planned approach to occupational health and safety are threefold:

1. *Economic*: Workplace injuries and poor employee health are a significant cost to industry. A preventative approach to health and safety can reduce these costs, as well as leading to less tangible cost benefits, such as the increased production caused by a more

Why Plan for Occupational Health and Safety? 3

harmonious and fulfilled workforce. The economic impact of health and safety problems is discussed at ¶103.

2. *Human*: The fact that employees may be injured or killed is an incentive to take all reasonable steps to ensure the safest possible workplace.

3. *Legal*: Occupational health and safety legislation places a duty on employers to ensure the health and safety of all their employees. More specific laws then regulate the use of certain equipment and substances, as well as factors such as noise levels, lighting, ventilation, first aid facilities and storage of dangerous chemicals. Regardless of whether there is specific legislation on an issue, the general duty to ensure the safety of employees will apply, and a failure to do so can result in fines or a prosecution against the employer. The legislative framework underpinning occupational health and safety is set out in Chapter 2.

> The maximum benefit from occupational health and safety can only be realised when it is identified as being our individual and personal responsibility, to which we pledge our full support and cooperation in an all out effort to achieve this common goal.

The purpose of this chapter is to indicate the magnitude of the health and safety problem for organisations to show why a planned approach to health and safety is essential. This introduction includes explanations of the role of management and employees in a planned health and safety program and the aims of such a program.

¶102 Measuring the problem

The International Labour Organisation (ILO) estimates that over one million people worldwide die each year in work-related accidents, an average of one person every two minutes. This is more than the worldwide yearly death toll from road accidents, war, violence and HIV/AIDS, according to Dr Jukka Takala, Chief of the ILO's OHS Branch, in his address to the April 1999 World Congress on Occupational Safety and Health. Although work-related injury and death rates are lower in advanced industrialised countries such as Australia than in many other parts of the world, there is no doubt that work-related illhealth and injury are a severe drain on the economy as a whole, as well as on individual organisations.

The scope of the impact

It is difficult to assess the magnitude of the problem accurately. In Australia, the National Occupational Health and Safety Commission

(NOHSC) provides some statistics on the scope of occupational injuries in this country. In 1996/97, there were 404 compensable workplace fatalities in Australia (excluding the Australian Capital Territory), and 121,666 workers compensation cases which resulted in a fatality, permanent disability or a temporary disability resulting in an absence from work of more than one week (excluding Victoria and the Australian Capital Territory). The National Commission estimates that in addition to this, there are about 2,200 deaths a year due to workplace exposure to hazardous substances such as asbestos and harmful chemicals.

Data based on workers compensation figures understate the magnitude of the problem, because many cases of work-related injury or disease never result in a claim for compensation. Examples of those not covered by workers compensation insurance include sole traders, and people whose work-related diseases are never recognised as being wholly or partly attributable to workplace exposures. Other hidden costs such as rehabilitation centres and other hospitalisation costs also add to the total burden on the community.

In the light of Australian industry's strivings for international competitiveness, there is substantial potential for cost reductions in this area. The restructuring of State compensation systems in recent years (primarily a reduction in the access to common law claims and an increased emphasis on rehabilitation) has resulted in reduced workers compensation premiums, but more remains to be done.

The human factor

Besides the "drama" of death or personal injury, there is a hidden incidence of occupational disease, which in many cases is a slow deterioration going unnoticed in the early stages. Examples include occupational hearing loss, dust diseases and occupational cancer. It may be only after a time delay of five years or more that the existence of such diseases is established, by which time their effects may have become irreversible.

The social costs of work-related injury (to the affected employee, his/her family and the community) can be less obvious than the economic impact. An employee whose injury disallows continuing employment becomes to some extent invisible, and once the individual matter of compensation is settled the incident may no longer be an influence for change at that workplace or others.

¶102

Why Plan for Occupational Health and Safety?

¶103 Indirect costs

There can be no dispute that injuries, ill health and deaths of workers represent the greatest insurance costs to an organisation. But on top of these direct insurance costs and the basic human considerations are other losses of which management should be particularly aware. These include indirect costs such as loss of productivity, loss of morale, labour turnover, absenteeism, industrial disputation over working conditions, and an impaired public image which may be reflected in loss of orders/accounts.

An organisation with an active, ongoing, consultation-based health and safety program will have reduced the risk of accidents occurring and have efficient procedures to deal with the human or technical emergencies that become part of a work accident. Money, time and, most importantly, lives can be saved. The data gathered through accident investigation can be fed back into the overall management administration to help control future incidents.

An American study by Frank E Bird Jnr, Director of Safety and Engineering Services for the Insurance Company of North America, produced an accident ratio "pyramid" showing the relative occurrence of different types of accident. An example is illustrated below.

Accident ratio study

ACCIDENT RATIO STUDY

Level	Count	Category	Description
Top	1	Serious or Disabling	Includes disabling and serious injuries
	10	Minor Injuries	Any reported injury less than serious
	30	Property Damage Accidents	All types
Base	600	Accidents with no Visible Injury or Damage	(Critical incidents)

6 Planning Occupational Health & Safety

The auditing of accident cost

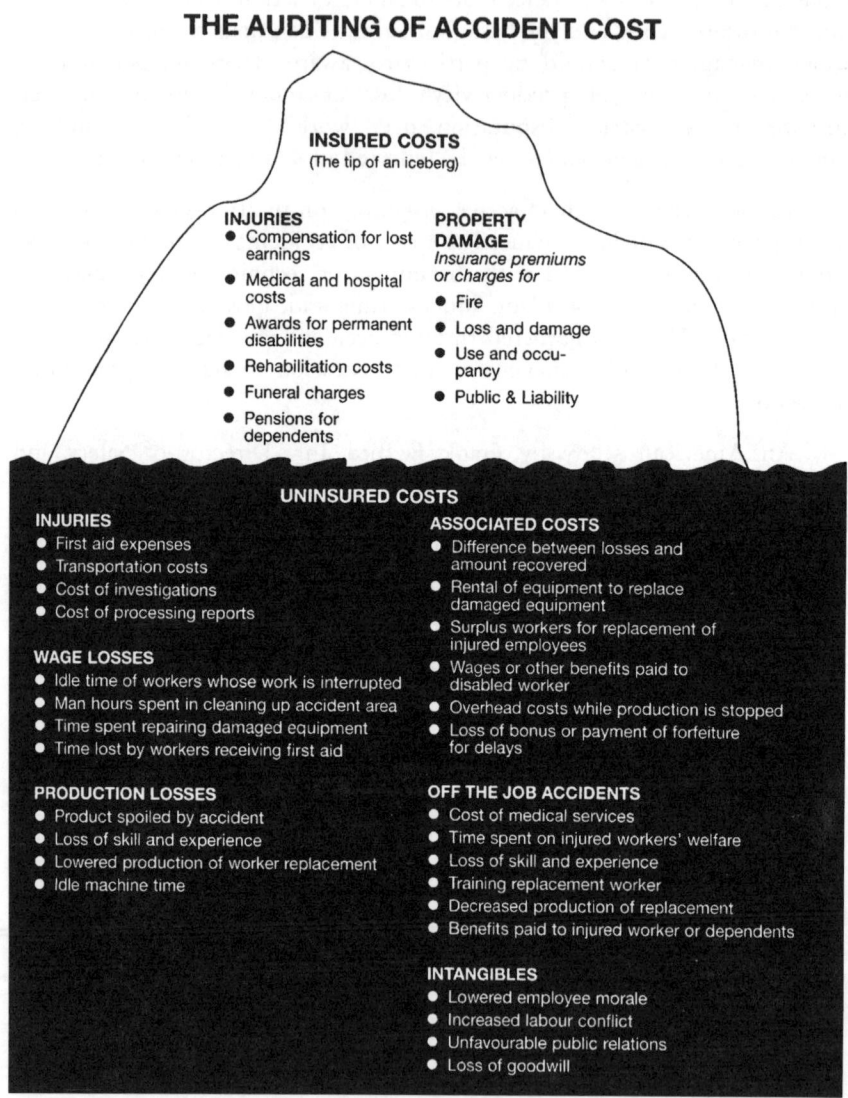

The magnitude of the indirect or hidden costs to management is illustrated by the previous diagram, which divides costs into "insured" and "uninsured" categories. Note that, although the "uninsured" costs are in

¶103

many instances difficult to calculate, their potential is to create a total cost several times that of the total insured costs. (The diagram is reproduced with the permission of the National Safety Council of Australia.)

Accidents may also have harmful effects beyond the workplace on the general environment. For example, if an oil barge discharges oil into a river, there may be damage to the foreshores, other boats, wildlife and nearby homes. Pollution from chimney discharge may affect nearby residents, cars parked in the street, etc. Management should also be aware of the implications of public liability (persons other than employees who become injured) and product liability (adverse consumer reaction, loss of sales, cost of recalls, etc).

¶104 Changes

The "cost" realisations referred to at ¶103 have prompted changes in the approach to the practice of occupational health and safety in Australia. These changes emphasise practical means to achieve injury prevention and to encourage health maintenance and enhancement at the workplace. The changes include the following:

1. The traditional "minimum standards"/compliance approach to the formulation of the law on health and safety is an inheritance from an earlier, more static, work environment and is ill suited to keeping pace with the rapid and uneven adoption of new work practices initiated by technological change. This has given way to a "performance-based"/ self-regulated approach which emphasises the employer's duty of care and a process of risk management (detailed in Chapter 3), as well as consultation with employees (outlined in Chapter 2). The term "performance-based" as it is applied to the modern style of health and safety legislation draws on the notion that any system of protecting workers from work-related injury or illness is only as good as its performance — that is, the outcome is of primary importance, rather than compliance with minimum standards.

2. Increasingly, codes of practice provide detailed instructions on how certain industries can operate safely, how to avoid certain common injuries (such as Occupational Overuse Syndrome, manual handling injuries and occupational hearing loss) or how to handle certain dangerous substances (such as asbestos, vinyl chloride or synthetic mineral fibres). National codes of practice produced by the National Occupational Health and Safety Commission are merely advisory, whilst some States have chosen to adopt codes of practice in their legislation.

3. All States and the Northern Territory have revamped their workers compensation schemes, which continue to be volatile in their attempts to deal with spiralling workers compensation costs.
4. Historically, occupational health and safety has been regulated by separate legislation in each State and Territory, with the result that in Australia there are nine separate sets of health and safety laws (for the six States, two Territories and the Commonwealth). The establishment of the National Occupational Health and Safety Commission has created a gradual movement towards more uniformity of requirements. The Commission includes representatives of Federal and State Governments, employers and employees. It aims to support an effective national infrastructure for the promotion of occupational health and safety, and accordingly facilitates the development of a nationally consistent framework of standards, as well as providing national data and practical guidance for industry.

¶105 Management's role

The main emphasis of this book is that the dangers of the workplace cannot be dealt with haphazardly; they must be prevented, or at least contained, through the basic functions of risk management, planning, organising, staffing and controlling.

Management's role in this process should be clearly seen and demonstrated. The employment of trained occupational health and safety personnel, the establishment of health and safety representatives and/or committees and cooperation with these, as well as the provision of adequate financial backing for their function, will, of course, be management's responsibility. Most importantly, management at all organisational levels needs to demonstrate by personal example and positive attitudes, in addition to its financial support, a commitment to high standards of health and safety for all employees wherever they may be employed. The standard of care demanded of an employer is high and one owed to individual workers rather than to workers in general.

If workers are not made aware of the reasons for the implementation of a health and safety program or do not see management staff complying with it, they will become suspicious of the motives behind such a program. They may resent any interference with their established work patterns, and they may not report certain accidents or incidents if they feel they are being monitored or that they could be blamed or punished for reporting them. Therefore they must be convinced that change is essential before they will accept it.

Workers should be involved in the planning and implementation of health and safety programs (see ¶106). Yet it is management that must

Why Plan for Occupational Health and Safety?

accept final responsibility in regard to its workers, its shareholders, its legal requirements, and the community at large. It is only on the basis of management responsibility and accountability that occupational safety and health programs will succeed.

Strategic planning versus operational planning

Management will be concerned with ensuring that the organisation's overall planning (strategic planning) incorporates health and safety. This will involve giving health and safety a high priority, staffing and financing the health and safety function, and determining what safety issues should receive priority.

In relation to health and safety, operational planning (which is concerned with shorter term matters) will be carried out by the health and safety officers, supervisors or the human resources department (see ¶107).

¶106 Employees' role — developing a consultative approach

As has already been noted, the modern approach to occupational health and safety emphasises consultation between employers and employees. The usual method of achieving this is through health and safety representatives and/or a health and safety committee. In all States and Territories, there are legislative requirements to elect/appoint either representatives or a committee (see ¶204).

Health and safety representatives are generally elected by the employees, with the workplace being broken down into locations/units, and each location/unit electing one representative. Elected representatives will generally be more acceptable to employees than appointed representatives. Health and safety committees may be made up of elected representatives and appointed employer representatives. In each State, the legislation provides some framework as to how the representatives and/or committee members must be selected (see ¶204).

The functions of representatives and committee members are discussed at ¶507 and ¶508.

In addition to consultation through representatives and committees, employees should be made aware of all health and safety initiatives at the workplace through meetings, circulars and supervisor communication. It may be worthwhile to obtain employee input into the design of such programs. Employees should also be made aware of the results of all health and safety programs.

¶106

Advantages of a consultative approach

Involving workers in health and safety will have benefits in terms of both the information that will be available to the health and safety program and the commitment of the workers to the program.

With respect to the additional information workers can provide, they are the closest to the work process and therefore more aware of what methods, procedures and practices apply, which machines have certain problems or peculiarities, the types of minor incidents which occur frequently and the short cuts which are practised to complete jobs on time. Information such as this is invaluable to a health and safety program. Workers are also best placed to monitor a program after it is introduced.

Equally invaluable to the success of any program is the cooperation and commitment of the workers. These can be enhanced by consultation on workplace safety issues. Genuine consultation will demonstrate to workers that the program is a real attempt to improve their safety, rather than an exercise in minimum standards compliance. Workers who become involved in the consultation process (that is, those who become representatives or committee members) can be expected to greatly increase the awareness of their own safety, and the safety of their immediate section.

¶107 The role of the human resources department/health and safety department

It has already been noted that the commitment to health and safety must come from the top of an organisation. However, the day-to-day resourcing of the health and safety function may be carried out with the assistance of a human resources department. In large organisations a specific health and safety department may be constituted.

Such day-to-day resourcing may involve a wide range of functions, including assistance with the following:

- identifying hazards to health and safety, and assessing the risks arising from those hazards;
- controlling the risks to health and safety, and monitoring the success of the control measures;
- implementing specific programs;
- preparing feedback to management on the effectiveness of programs;
- liaising with health and safety professionals, and/or medical professionals;
- administering health and safety committee meetings or meetings between representatives and management;

Why Plan for Occupational Health and Safety?

- recording all unsafe incidents and recommending changes arising out of such incidents;
- performing regular inspections or arranging for professionals and worker representatives to perform inspections;
- maintaining first aid equipment and a first aid room;
- monitoring the condition of any safety equipment; and
- compiling and keeping up to date all safety information, including Material Safety Data Sheets.

Those involved in managing health and safety may encounter a certain resistance to new safety practices, perhaps as part of a general resistance to change. Disputes can best be avoided through proper consultation and education.

¶108 Aims of an occupational health and safety program

The aim of an occupational health and safety program is to identify and remove or control the causes of accidents and health hazards at the workplace. This will involve an assessment of the current safety and health position at the workplace, correction of any hazards, development and documentation of safe work methods to a satisfactory standard with the establishment of health and safety education, promotion and training programs.

The goals of a health and safety program will be best achieved through the involvement of both workers and management in some form of health and safety organisation (such as health and safety committees or health and safety representatives) and through the employment of health and safety personnel (such as occupational health physicians and health and safety officers). Such staffing matters are discussed in Chapter 5. Setting up a health and safety program is discussed in Chapter 7.

¶109 Some definitions

Before dealing with the process of developing an occupational safety and health program, it would be helpful to clarify a few of the terms used in the occupational health and safety field.

Occupational health

First of all, the basic concept of "occupational health" needs to be considered. A joint International Labour Organisation (ILO) and World Health Organisation (WHO) Committee on Occupational Health (1950) described the objective behind occupational health as follows:

> Occupational health should aim at: the promotion and maintenance of the highest degree of physical, mental and social well-being of workers

in all occupations; the prevention among workers of departures from health caused by their working conditions; the protection of workers in their employment from risks resulting from factors adverse to health; the placing of and maintenance of the worker in an occupational environment adapted to his physiological and psychological equipment and, to summarise, the adaption of work to man and of each man to his job.

The concept of "occupational health" is, then, a positive one, which focuses on the need for maintenance of good health and wellbeing. Of course, occupational health and safety programs will deal with the treatment of ill health as well, but prevention should be the primary motive in any program.

Occupational health and safety matters are defined under the *National Occupational Health and Safety Commission Act 1985* (the Act establishing the Commission) as including:

"(a) the physiological and psychological needs and well-being of persons engaged in occupations;

(b) work-related death;

(c) work-related trauma;

(d) the prevention of work-related death or work-related trauma;

(e) the protection of persons from, or from risk of, work-related death or work-related trauma;

(f) the rehabilitation and re-training of persons who have suffered work-related trauma."

Ergonomics

Planning a safe, healthy and comfortable working environment necessitates a basic understanding of the concept of ergonomics. Ergonomics may be defined as the scientific study of the physical relationship between people, the equipment they use and their general environment. As an applied science, ergonomics is involved with the design of equipment and working environments that enable the best use of human capabilities without exceeding human limitations. (In the United States ergonomics is usually referred to as human factors engineering.)

To some extent the term "ergonomics" has become debased through an insufficient understanding of the problems to which it is applied and the mislabelling of half-considered "solutions" as ergonomic. A striking example of this can be seen in the often costly purchase of "ergonomically designed" furniture as a solution to problems encountered with the introduction of computer technology. While the design of some furniture can contribute to postural problems, more appropriately designed furniture

Why Plan for Occupational Health and Safety?

does not prevent a person adopting an inefficient and potentially damaging working posture, when not used properly.

Ergonomics cannot rightly be considered as a single subject but rather as a *principle* that should be involved in all aspects of work design and planning. For instance, ergonomic considerations are involved in subjects as diverse as the design of hand tools, the positioning of machine controls and the layout of workplaces. It also includes highly specialised studies such as the application of psychology to the ability of the brain to absorb and use information from display screens or the capacity of the body to work under extreme conditions of temperature.

Accidents and incidents

A workplace accident may be described as any event arising out of employment which results in a work injury or damage to property, or creates the possibility of injury or damage.

Near misses (incidents) and property damage should be taken into account in any accident prevention program as these events give warning that there is something in the work process that needs to be investigated and possibly changed before a serious accident occurs, as illustrated in the Accident Ratio Study at ¶103.

¶110 Scope of this book

Planning Occupational Health & Safety is written against a background of increasing emphasis on risk management in occupational health and safety.

Where relevant, reference is made to the impact of specific legislation, but generally the approach taken in this book is not legalistic but that of a guide to planning. Subsequent chapters of this book outline and discuss the issues and procedures involved in administering the safety and health function of an individual workplace.

Topics covered include assessing the current workplace, risk management, the health and safety audit, staffing issues, setting up a program or service, accident prevention, and evaluating a program, as well as an introduction to particular health problems.

Chapter 2

Overview of Legislation

Purpose of this chapter	¶201
Modern legislative approach	¶202
Summary of occupational health and safety legislation	¶203
Legislation requiring consultation with employees	¶204
Enforcement	¶205
Codes and standards	¶206
Common law liability	¶207
Workers compensation	¶208

¶201 Purpose of this chapter

As was mentioned in the previous chapter, Australian occupational health and safety has moved away from a minimum standards compliance approach in favour of a performance-based, consultative approach grounded in the principles of duty of care and risk management. The purpose of this chapter is, therefore, not to provide standards for management to comply with but to set out the legislative framework within which occupational health and safety operates.

Summary of types of legislation

There are various forms of legislative provisions relevant to occupational health and safety. The following summary is expanded upon in the remaining paragraphs of this chapter. Note that liability imposed by the common law exists in addition to the legislative provisions.

1. *Statutes*: The States and Territories are responsible for the law and its administration in the area of occupational health and safety. Each State and Territory has its own (variously titled) Health and Safety Act,

largely general in its nature, which requires employers to ensure the health and safety of their employees. The Acts also create administrative bodies and empower the government to make regulations covering specific areas. Another feature of these Acts is the requirement to establish health and safety representatives/committees.

Each State also has its own workers compensation legislation, which provides for those injured in the course of their employment. Requirements to rehabilitate workers may be contained within workers compensation legislation.

There are also statutes in each State setting out safety requirements for specific industries or areas (for example, the coal industry and dangerous goods).

2. *Regulations* are detailed legislative provisions specifying requirements, duties and procedures in specific areas. Some regulations will affect all workplaces while others apply to particular industries.

3. *Codes and standards*: Codes of practice are not legislation, but have the backing of Government and act as guides to procedure in particular areas. Codes can be used in determining whether an employer has breached its duty to ensure the safety of employees.

   Standards are similarly not legislation (although there are many instances where an Australian Standard has been incorporated into regulations). They are published by Standards Australia (see ¶1002) after consultation with concerned groups, and are accepted as authoritative guides to good practice. The National Occupational Health and Safety Commission has assumed responsibility for developing national standards relating to occupational health and safety.

4. *Industrial awards and agreements* are made pursuant to federal or State industrial legislation, and are enforceable at law. They often contain provisions relevant to health and safety, such as protective clothing, first aid facilities and regular rest breaks.

   Industrial awards and enterprise bargaining agreements are not considered in any detail in this book. Note that in theory industrial awards and enterprise bargains could deal specifically with health and safety and, where this occurred, they would have to be taken into account, as well as State occupational health and safety legislation, in relation to that industry or workplace.

Overview of Legislation

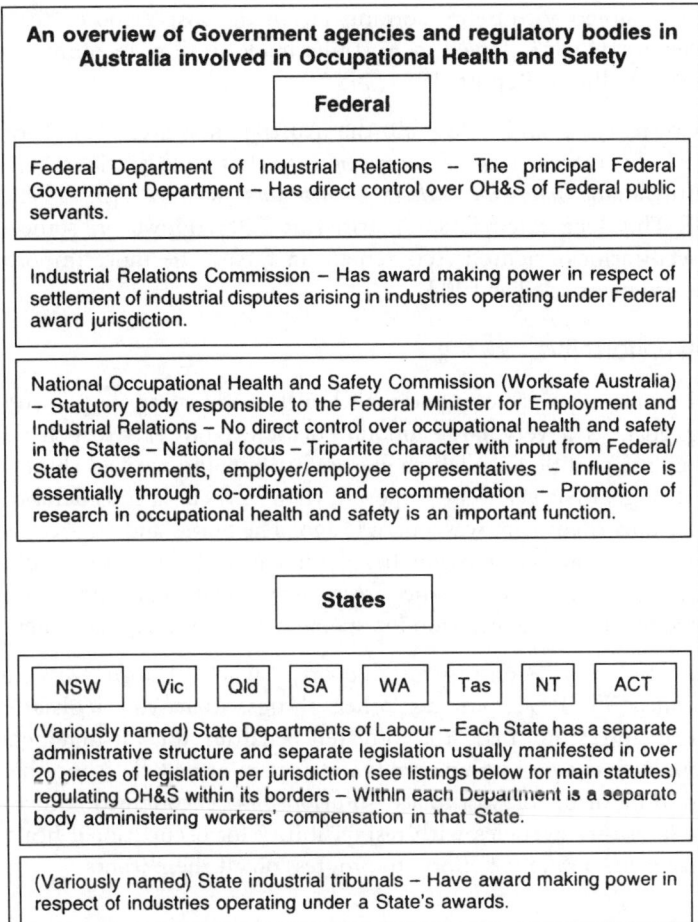

¶202 Modern legislative approach

In the late 1960s occupational health and safety legislation in Australia was based solely on the specification of standards. The legislation was directly descended from the earliest English Factories Acts. A realisation that this type of legislation did not, in fact, lead to a safer workplace has led to the introduction of a different style of legislation.

Impetus for the new style of legislation came from the British Robens Inquiry into Safety and Health at Work, which released its report in 1972. The *Robens Report* recommended a change in direction for occupational health and safety laws. The new legislation should be based on the expression of general duties of care and provisions for the participation of employees, coupled with better administration and inspection of health and safety. Reports which followed in Australia made similiar recommendations (notably the Williams Report, 1981, NSW).

Between 1972 and 1989 all the Australian States and Territories introduced the newer style of legislation, with the forerunner being the South Australian *Industrial Safety, Health and Welfare Act, 1972* (now repealed). This legislation is summarised at ¶203. However, some of the older, specifications-oriented Acts remain in force. The more important of these are also mentioned at ¶203.

National uniformity

National uniformity in occupational health and safety legislation is a daunting goal, as a very large amount of State and Territory legislation presents points of difference which are gradually being worked through and aligned. The goal has yet to be achieved, though there is considerably more uniformity now than there was a decade ago. The States and Territories have progressively adopted legislation based on national standards and model regulations, for example in the areas of manual handling, hazardous substances, plant and certification for operators of industrial equipment.

The Industry Commission report, *Work Health and Safety*, recommended in 1995, among other things: that the legislation be streamlined and simplified; that greater penalties for infringement and greater enforcement be implemented; and that informed choice, rather than regulated imposition of workplace solutions, be encouraged. Since then, State and Territory agencies with responsibility for occupational health and safety have made a range of efforts to progress on all these fronts.

Risk management

Experience has shown that a legislative approach relying on the prescription of minimum standards and enforcement by inspectors is not always effective. It neglects many of the risk factors for work-related injury and illness, and does not provide employers with a clear process for achieving good safety and health standards. In order to address the hazards in a workplace, risk management needs to be carried out.

Definition

Risk management involves a detailed and systematic examination of any activity, location or operational system to identify risks, understand the likelihood and potential consequences of the risks, and to review the possible approaches to controlling the risks. Successful risk control can include outcomes such as improved safety, health, production, environmental

Overview of Legislation

protection and community acceptance. In the context of occupational health and safety, the first stage of the risk management process is usually known as "hazard" identification, rather than "risk" identification.

Risk management and legislation

The modern approach in drafting legislation and codes of practice for occupational health and safety is to require that risk management underpins any implementation of health and safety measures. See Chapter 3 for details of the risk management process.

¶203 Summary of occupational health and safety legislation

The following provides an overview of the principal health and safety legislation in each jurisdiction. Other relevant Acts are simply mentioned.

Note that the principal Acts do not provide exact specifications for employers to follow. Their purpose is to provide general standards for employers to meet, and to allow prosecutions and/or recovery of damages when those standards are not met. Any specifications which exist are set out in regulations or codes of practice made pursuant to the legislation, or in other, more specific legislation.

The principal Acts differ widely in their terms of coverage and provisions. It would be advisable for employers and employee representatives to maintain up-to-date copies of the Acts applicable to their workplace. The loose leaf reporting service *Australian Occupational Health & Safety Law* (CCH Australia Limited) contains the full text of all principal legislation, with extracts from other legislation.

The following description of legislation is correct as at 1 May 2000. Requirements in the legislation to establish health and safety representatives or committees are discussed separately at ¶204. Enforcement of the legislation is discussed at ¶205.

Federal

The Commonwealth has no constitutional power to legislate generally in the area of health and safety. The *Occupational Health and Safety (Commonwealth Employment) Act 1991* regulates the health and safety of federal public servants. It is administered by Comcare Australia.

The Act places general duties of care on employers (the Commonwealth or a Commonwealth authority), employees and manufacturers and suppliers in much the same way as the State Acts discussed below. The Act also provides for the selection of health and safety representatives in Commonwealth workplaces and, in certain circumstances where the number of employees at the workplace is at least 50, for the establishment of health and safety committees.

Other federal legislation relevant to occupational health and safety includes:

- *National Occupational Health and Safety Commission Act 1985*: established the National Occupational Health and Safety Commission; and

- *Industrial Chemicals (Notification and Assessment) Act 1989*: establishes a scheme whereby all new chemicals imported into or manufactured in Australia are assessed (known as the National Industrial Chemical Notification and Assessment Scheme (NICNAS)).

New South Wales

The principal Act is the *Occupational Health and Safety Act 1983*, administered by the WorkCover Authority of New South Wales. The Act places a general duty on employers to ensure the health, safety and welfare at work of all their employees (sec 15(1)). This is extended to persons other than employees at the workplace (sec 16). There is also a duty of care placed on manufacturers and suppliers of plant and substances (sec 18), and on employees to take reasonable care for the health and safety of other persons at the workplace (sec 19).

Health and safety inspectors are given the power to inspect any workplace (sec 31A). Regulations made under the Act allow an inspector to issue improvement notices or prohibition notices requiring an employer to remedy a particular safety risk, to issue on-the-spot fines or to initiate prosecutions.

The Act provides for the establishment of workplace health and safety committees (see ¶204).

WorkCover is given the power to formulate industry codes of practice (sec 44A). Such codes are not in themselves legislation but they may be used in proceedings under the Act to establish that a person failed to comply with his/her duty of care (see ¶206).

Regulations made under the Act prescribe safety standards in a number of areas of health and safety (for example, first aid equipment and facilities, asbestos removal, notification of accidents, and so on). The Regulations under the NSW *Occupational Health and Safety Act 1983* are currently under review, and it is expected that they will be consolidated under a new OHS

Overview of Legislation

Regulation which maintains many existing provisions and encourages a risk management approach.

Other relevant legislation includes:

- *Factories, Shops and Industries Act 1962*: regulations made under the Act prescribe minimum safety standards for a large number of industrial processes;
- *Construction Safety Act 1912* and associated regulations;
- *Coal Mines Regulation Act 1982*;
- *Dangerous Goods Act 1975*; and
- *Electricity Safety Act 1945*.

Victoria

The *Occupational Health and Safety Act 1985* is the principal legislation in Victoria. Occupational health and safety in Victoria is administered by the Victorian WorkCover Authority.

The Act establishes various general duties of care including: the duty of employers to provide and maintain a safe working environment for employees (sec 21); the duty of employers and self-employed persons to ensure that others are not exposed to health and safety risks arising from their undertaking (sec 22); the duty of occupiers to ensure that the workplace, and access to the workplace, is safe (sec 23); the duty of manufacturers, importers and suppliers of plant and substances to ensure that the plant or substances are safe for use (sec 24); and the duty of employees to take care of their own safety and the safety of others at the workplace (sec 25).

The Act also provides a mechanism for dealing with health and safety disputes (sec 26). The employer and the health and safety representative should attempt to resolve the dispute, with either having the power to direct that work shall cease where the threat is immediate, or to require an inspector appointed under the Act to attend the workplace. Where work ceases because of a safety threat, the employer is entitled to assign the employees to suitable alternative duties.

The Act contains provisions granting inspectors the power to enter and examine any workplace (sec 39). Inspectors have the power to issue improvement notices where there is a contravention of the Act or regulations (sec 43), or prohibition notices in more serious cases (sec 44). It is an offence to fail to comply with either type of notice.

¶203

The *Occupational Health and Safety Act* deals with the establishment of health and safety representatives and committees, and provides representatives with the power to issue improvement notices to the employer. This is discussed in more detail at ¶204.

A number of regulations, dealing with specific topics, have been issued pursuant to the Act. Requirements regarding the notification of accidents and the keeping of accident records are contained in the *Occupational Health and Safety (General Safety) Regulations 1986*. The Act also allows for the approval of codes of practice "the purpose of providing practical guidance to employers, self-employed people, employees, occupiers, designers, manufacturers, importers, suppliers or any other person who may be placed under an obligation by or under this Act" (sec 55(1)). A number of such codes have been approved (see ¶206). The codes do not have legislative force, but can be used in proceedings under the Act to help establish a breach of the Act or regulations.

Other relevant legislation includes:

- *Labour and Industry Act 1958*: provides certain minimum standards for factories and other workplaces; regulations made pursuant to the Act cover a wide range of matters;
- *Equipment (Public Safety) Act 1994*;
- *Building Act 1993*: provides for building standards;
- *Dangerous Goods Act 1985*; and
- *Health Act 1958*: regulations made under the Act cover, among other things, confined spaces, fire prevention, harmful gases and hearing conservation.

Queensland

The *Workplace Health and Safety Act 1995* is the principal legislation in Queensland. The administering body is the Workplace Health and Safety section of the Department of Training and Industrial Relations. The Act's objective is to ensure freedom from disease and injury, and the risk of disease and injury, caused by workplaces, workplace activities or specified high-risk plant (sec 7).

Obligations for ensuring health and safety fall on employers, self-employed persons, persons in control of workplaces, principal contractors, designers, manufacturers, importers and suppliers of substances, and owners of high-risk plant (sec 23). Each of these must ensure the health and safety of both themselves and other people in the workplace, that plant and

Overview of Legislation

substances used at the workplace are safe, and that risks are minimised (sec 28-36).

The employer, self-employed person or principal contractor must record any injuries, illnesses or dangerous events which occur (*Workplace Health and Safety Regulation 1995* (sec 15)).

Inspectors appointed for the purposes of the Act are given the power to enter any premises and carry out such examinations as are necessary to ensure compliance with the Act (sec 104). Where an inspector is of the opinion that a person is contravening the Act, the inspector may issue an improvement notice, and the failure to comply with such a notice is an offence under the Act (sec 117). In more extreme circumstances the inspector may issue a prohibition notice, requiring certain work or the use of a certain substance to stop until the safety risk is remedied (sec 118).

The *Workplace Health and Safety Act* (sec 92) requires the appointment by the employer of a "qualified person" as health and safety officer in all workplaces with 30 or more workers. To become suitably qualified, a person must apply for a certificate of authority to the chief executive of the Workplace Health and Safety Council. The health and safety officer tells the employer or principal contractor about the overall state of health and safety in the workplace; conducts inspections to identify hazards or unsafe conditions and practices and reports these to the employer; reports any work-caused illness, injury or dangerous event to the employer; establishes educational programs in workplace health and safety; investigates or assists in the investigation of work-caused illnesses and dangerous events; and assist inspectors with their duties (sec 96).

The Act also allows the possibility of the establishment of health and safety representatives and committees (see ¶204).

Regulations made pursuant to the Act cover a wide range of matters. In addition, the Act provides for the creation of compliance standards (which are subordinate legislation) to minimise exposure to risk (sec 38), and advisory standards (sec 40) to identify and manage exposure to risk at the workplace. Advisory standards do not have the force of law, but can be used as evidence in proceedings related to alleged contravention of the Act (sec 42).

Other legislation relevant to health and safety includes:

- *Building Act 1975*: prescribes standards relating to building construction, including fire safety requirements; and
- *Mines Regulation Act 1964*.

South Australia

The principal legislation is the *Occupational Health, Safety and Welfare Act 1986*. It is administered by the WorkCover Corporation of South Australia. The Act establishes the Occupational Health, Safety and Welfare Advisory Committee (sec 7), a committee of nine members advising the government on occupational health and safety policy, administering and enforcing standards, regulations and codes of practice and advising on their implementation and revision, among other things (sec 8).

The Act places a general duty on employers to ensure the safety of employees, in particular to provide a safe working environment, a safe system of work, and safe plant and substances (sec 19). Further, employers are required to prepare and maintain, in consultation with employees and their representatives, policies relating to health and safety in the workplace (sec 20). Written statements setting out the practices and procedures adopted to protect the safety of the employees must be prepared and brought to the attention of the employees.

The Act also places general duties of care on employees (sec 21), employers in relation to persons other than employees at the workplace (sec 22), occupiers (sec 23), designers and owners of buildings (sec 23A), and manufacturers and suppliers of plant and substances (sec 24).

Corporations must appoint a responsible officer to ensure that the corporation complies with its obligations under the Act (sec 61). The responsible officer must reside in the State and must be the chief executive officer or a member of the governing body or, where that is not possible, a senior executive officer or officer of the corporation.

The Act provides for the election of health and safety representatives and, in certain circumstances, for the establishment of health and safety committees (see ¶204).

The power of an inspector to enter and examine any workplace is set out in sec 38 of the Act. Inspectors are given the power to issue improvement notices (sec 39) or, where the inspector is of the opinion that there is an immediate risk to health and safety, a prohibition notice (sec 40). Failure to comply with either type of notice is an offence and the Act provides harsh penalties.

Regulations made under the Act cover a large number of specific activities and industries, including hazard identification and risk assessment, hazardous substances and plant safety. The Act also makes provision for codes of practice (sec 63), which can be used in proceedings under the Act to establish that the employer did not exercise the required standard of care. A number of such codes presently exist (see ¶206).

Other relevant legislation includes:

- *Dangerous Substances Act 1979.*

Western Australia

The principal legislation is the *Occupational Safety and Health Act 1984*, which establishes the WorkSafe Western Australia Commission. The Act requires employers to provide and maintain a working environment in which employees are not exposed to hazards (sec 19). This includes the provision of safe workplaces, safe plant and safe systems of work, providing information and training to enable employees to carry out their work safely, and consulting and cooperating with health and safety representatives. Employers are required to notify the Commission whenever there is an accident at the workplace which results in death or certain serious injuries as prescribed in the regulations (sec 19(3)).

Employees are required to take reasonable care of their own safety and the safety of others at the workplace (sec 20). General duties of care are also placed on employers and self-employed persons in relation to persons at the workplace other than employees (sec 21), and manufacturers and suppliers of plant (sec 23). Manufacturers and suppliers of substances are required to ensure that adequate toxicological data is provided with the substance (sec 23(3)).

The Act provides for meetings between the employer and the health and safety representative(s)/committee or the employees in order to resolve health and safety disputes (sec 24; see also reg 202). Where a dispute cannot be resolved an inspector may be notified and may attend the workplace and take such action as he/she sees fit, including an order to cease work (sec 25). An employee with reasonable grounds for believing that there is an imminent and serious safety threat may cease work, although he/she must immediately notify the employer and the health and safety representative (sec 26). Such employees will still be entitled to their usual pay, although the employer is entitled to require them to perform reasonable alternative work.

The Act provides for the election of health and safety representatives and, in certain circumstances, for the establishment of health and safety committees (see ¶204).

Section 43 governs the powers of inspectors to enter and examine any workplace. Inspectors are also given the power to issue improvement notices (sec 48) or prohibition notices (sec 49). Such notices may include directions as to the measures an employer must take, and the failure to comply with a notice is an offence.

The Act allows for the Minister to approve codes of practice for "the purpose of providing practical guidance to employers, self-employed persons [and] employees" (sec 57). Such codes are not given any legislative force or evidentiary weight in proceedings under the Act.

The *Occupational Safety and Health Regulations 1988* cover a large range of matters, including noise protection, workplace amenities, fire safety equipment, use of plant, first aid facilities and hazardous substances.

Other relevant legislation includes:

- *Explosives and Dangerous Goods Act 1961*;
- *Mines Safety and Inspection Act 1994*; and
- *Radiation Safety Act 1975*.

Tasmania

The principal legislation in Tasmania is the *Workplace Health and Safety Act 1995*, which is administered by Workplace Standards Tasmania. The Act relates the duties and obligations of employers, responsible officers, self-employed persons, designers, manufacturers, importers, suppliers and installers regarding workplace health and safety (sec 9-21). Codes of practice may be approved under the Act (sec 22).

Serious accidents and dangerous incidents are to be reported to an inspector by the quickest means available (sec 47). Regulations to the Act require a record of work injuries to be kept by the employer.

An inspector may direct the employer to remedy or alleviate dangerous circumstances in the workplace by a notice in writing (sec 38).

The Act also provides for the election of safety representatives, with further details relating to safety representatives being contained in the regulations (see ¶204). Regulations made pursuant to the Act cover a wide range of specific matters, including fire safety, noise protection and machine safety.

Other relevant legislation includes:

- *Dangerous Goods Act 1998*; and
- *Radiation Control Act 1997*.

Australian Capital Territory

The principal legislation in the Australian Capital Territory is the *Occupational Health and Safety Act 1989*, administered by the ACT WorkCover Authority. The Act establishes the Occupational Health and Safety Council, a nine-member council charged with promoting occupational health and safety in the Territory.

Employers in the Territory are obliged to take all reasonably practicable steps to protect the health, welfare and safety at work of employees (sec 27) and other persons at or near the workplace (sec 28). The Act also places general duties of care on persons who have control of workplaces (sec 29), employees (sec 30), self-employed persons (sec 31), manufacturers (sec 32), suppliers (sec 33) and those erecting or installing plant (sec 34).

The Act requires employers with 10 or more employees to establish designated work groups, the members of which elect health and safety representatives (see ¶204). Health and safety representatives have the power to issue improvement notices in certain situations (sec 51), or to order an employee to cease work where there is an immediate threat to the health or safety of the employee (sec 56).

The *Occupational Health and Safety Act 1989* also makes provision for inspectors. Inspectors have the power to issue improvement notices where they believe there is a contravention of the Act or the regulations (sec 76), or prohibition notices where they believe there is a risk of serious and imminent injury (sec 77).

Employers are required to give the Occupational Health and Safety Registrar notice of a death or injury at or near the workplace (sec 85). Records of such injuries must be kept (sec 86). Regulations and codes of practice may be made pursuant to the Act.

Other relevant legislation includes:

- *Machinery Act 1949*: regulations made pursuant to the Act cover the installation, operation and inspection of machinery; they include the *Boiler and Pressure Vessels Regulations*;
- *Building Act 1972*: relates to the safety of construction and demolition work;
- *Dangerous Goods Act 1984*; and
- *Smoke-free Areas (Enclosed Public Places) Act 1994*.

¶203

Northern Territory

The *Work Health Act 1986* deals with both occupational health and safety and workers compensation. The Act is administered by Northern Territory Work Health, which is part of the Department of Industries and Business.

The Act places a general duty on employers to provide and maintain, so far as is practicable, a workplace that is safe and without risk to the health of workers or others at the workplace (sec 29). General duties of care are also placed on occupiers (sec 30), self-employed persons (sec 30A), manufacturers, importers, suppliers, erectors, installers (sec 30B), owners (sec 30C) and workers (sec 31). Where there is an immediate risk of severe injury a worker may cease work in that area (sec 32), although the Act allows an employer to assign such a worker to alternative work.

The Act allows the Authority to appoint work health officers (sec 35), who have the power to issue improvement notices (sec 40) or prohibition notices (sec 41). Work health officers are also given the power to carry out investigations into health and safety matters (sec 37).

The *Work Health (Occupational Health and Safety) Regulations 1992* deal with hazardous substances, plant, construction work, the work environment and specific work practices. Codes of practice may be approved by the Minister. While such codes are not legislation, they may be used in proceedings under the Act to establish a breach of the Act.

Other legislation relevant to occupational health and safety includes:

- *Dangerous Goods Act 1980*;
- *Mines Management Act 1990*; and
- *Radiation (Safety Control) Act 1998*.

¶204   Legislation requiring consultation with employees

Legislation in each State and Territory provides for the establishment of occupational health and safety representatives and/or committees. This paragraph provides a summary of those requirements.

New South Wales

The New South Wales legislation refers only to health and safety committees and not representatives.

The *Occupational Health and Safety Act 1983* requires health and safety committees to be established in all workplaces where 20 or more persons are employed and a majority of those employees requests the establishment of a committee, or where the WorkCover Authority directs that a committee be

Overview of Legislation

formed (sec 23(1)). Committee members have the right to inspect the workplace and to obtain information from the employer relating to the workplace (sec 25(1)). They also have the right to request a health and safety inspector to conduct an investigation at the workplace (sec 24). Committee members are to be provided with training to assist them to carry out their functions (sec 25(2)).

The *Occupational Health and Safety (Committees in Workplaces) Regulation 1999* contains additional provisions regarding the establishment of workplace committees, their membership and procedures, the powers and functions of their members and training requirements.

The functions of committees include to:

- assist in the development of appropriate recording systems of accidents and hazardous situations;
- assist in the development of a safe working environment and safe systems of work;
- assist in the formation of a policy;
- monitor the measures taken to ensure the proper use, maintenance and, if necessary, replacement of equipment designed to protect employees from hazardous situations; and
- make such recommendation to the employer as it thinks appropriate to ensure the health and safety of persons at that place of work.

Victoria

The *Occupational Health and Safety Act 1985* provides for the determination of "designated work groups". Section 29(1) allows an employee to request an employer to establish "designated work groups" at the workplace. The groupings may be formulated on bases such as the nature of the work or hazards at the workplace, or the geographic locations of workers. Each such group may then elect a health and safety representative.

Health and safety representatives are given the power to inspect the workplace at any time after giving reasonable notice to the employer, or immediately in the event of any accident or dangerous occurrence (sec 31(1)). They are also entitled to accompany an inspector during any inspection of the workplace, and to call in assistance from an expert whenever necessary. Employers are required to provide representatives with access to health and safety information and to consult with them on all proposed changes to the workplace. Representatives are entitled to time off

¶204

without loss of pay to attend training courses and to carry out their duties (sec 31(2)).

Further, health and safety representatives are given the power to issue provisional improvement notices (sec 33). It is an offence for an employer to fail to comply with such a notice, although the employer may require an inspector to attend the workplace to confirm or cancel the notice. There is also scope for representatives to consult with the employer over safety issues, and to order a cessation of work where there is an immediate health and safety threat to any employee (sec 26).

The Victorian Act makes some provision for the establishment of health and safety committees, although they are secondary to health and safety representatives. A representative is given the right to request the establishment of a joint health and safety committee (sec 31(1)). At least half the members of such a committee must be employees. The committee should meet at least every three months.

Queensland

The *Workplace Health and Safety Act 1995* requires employers or principal contractors to appoint a workplace health and safety officer for workplaces with 30 or more employees (sec 93-94). The officer is required to tell the employer or principal contractor about the overall state of health and safety at the workplace; conduct inspections and identify and report any hazards and unsafe practices; establish educational programs where necessary; investigate work-related injuries, illnesses and dangerous events; and report such occurrences to the employer or principal contractor (sec 96).

The Act allows employees to elect a health and safety representative (sec 67). They may elect more than one if the employer and employees agree. The representative is entitled to inspect the workplace; to be told by the employer of any work injury, illness or dangerous event; to report any such issues to the employer; to review such circumstances and make recommendations to the employer; to be consulted by the employer on any proposed change (to the workplace, plant or substances used) that affects health and safety; and to help resolve issues regarding workplace health and safety. The representative may also ask the employer to establish a workplace health and safety committee, and may be a member of that committee (sec 81).

The employer or principal contractor must establish a health and safety committee within 28 days of a request from the representative (sec 86). The chief executive of the Queensland Department of Training and Industrial Relations may direct the establishment of such a committee. The

¶204

composition of the committee is left to the employer's discretion, although it must comprise at least two people and include any health and safety officers and representatives. At least half must be workers elected by the employees (sec 87).

The committee's functions are assisting cooperation between the employer and employees to ensure workplace health and safety; considering appropriate training and education; informing workers about the standards and procedures affecting workplace health and safety; helping to resolve issues of health and safety; reviewing the circumstances surrounding workplace illnesses, injuries and dangerous events; and making recommendations to the employer (sec 90).

South Australia

The *Occupational Health, Safety and Welfare Act 1986* provides for work groups to elect a health and safety representative to represent the group (sec 27). The composition of work groups should be determined by consultation between the employer and the relevant employees, or a representative of them. Section 28 of the Act, along with the *Occupational Health, Safety and Welfare Regulations 1995*, sets out the procedure for the election of representatives. Such elections must be by secret ballot if any member of the work group so requests.

Representatives are able to inspect the workplace at any time after giving reasonable notice to the employer, or immediately in the event of an accident (sec 32(1)). They are entitled to bring in health and safety experts during such inspections. They are also entitled to accompany an inspector during any inspection of the workplace. Further, representatives may investigate any safety complaints made by members of the work group, and they are entitled to be present at any interview between the employer and an employee, or the inspector and employee, which concerns health and safety.

Employers are obliged to consult with health and safety representatives regarding any safety procedures at the workplace and any proposed changes to the workplace (sec 34). They must also provide representatives with access to safety information regarding the workplace, and allow a representative time off work, without loss of pay, in order to fulfil his/her functions and to attend approved training courses.

Health and safety representatives are also given the power to issue default notices where they believe the employer is contravening a provision of the Act (sec 35), and to direct an employee to cease work where they are of the opinion that there is an immediate threat to the health or safety of

¶204

that employee (sec 36). In either case, representatives must first consult with the employer.

The South Australian Act also makes provision for health and safety committees where the employer employs 20 or more employees. Such an employer must establish a committee if requested by a health and safety representative or at least five employees (sec 31 and *Occupational Health, Safety and Welfare Regulations 1995*, reg 6.2.1). The composition of the committee should be determined by agreement between the employer, the health and safety representative and any interested employees. At least half of the members of the committee must be employees (excluding managerial employees).

The functions of such committees include assisting in the formulation and dissemination of health and safety procedures, and assisting in the resolution of any safety disputes. The employer is obliged to consult with the committee over health and safety procedures to be used at the workplace, and over any proposed changes to the workplace.

Western Australia

The *Occupational Safety and Health Act 1984* and the *Occupational Safety and Health Regulations 1996* make provision for both health and safety representatives and committees.

Any employee may give notice to the employer requiring the election of a health and safety representative at the workplace (sec 29). Where an employer is given such a notice, it must consult with employee representatives on the number of health and safety representatives to be elected, the manner in which the election will be conducted, and the training to be given to such representatives (sec 30). The election is to be conducted by a person agreed upon between the employer and employees (sec 31).

A health and safety representative is able to inspect the workplace at times agreed with the employer (the employer must allow inspections at least every 30 days), or immediately in the event of an accident (sec 33). Other functions of the representative include reporting any safety hazards to the employer and liaising with the employer and employees on health and safety matters.

The employer is obliged to make available to a representative any information the employer has relating to health and safety at the workplace (sec 35). An employer is also obliged to confer with health and safety representatives about intended changes to the workplace, to permit representatives to be present at interviews between the employer and an employee concerning safety matters, and to notify representatives of any accidents or dangerous occurrences which occur at the workplace. The employer must allow a representative time off, without loss of pay, to perform his/her functions and to attend accredited training courses.

The health and safety representative may request the employer to establish a health and safety committee at any workplace of 10 or more employees (sec 36). The committee should be made up of the health and safety representative(s) for the workplace, persons elected by the employees and persons nominated by the employer (sec 38). The person(s) nominated by the employer shall be, or shall include, a person with the authority to put into practice such matters as the committee might resolve. At least half of the members must be either health and safety representatives or persons elected by the employees. Committees must meet at least every three months (sec 41).

The functions of a health and safety committee, as set out in sec 40, include facilitating cooperation between the employer and the employees relating to health and safety measures, and keeping itself informed of the standards of health and safety at comparable workplaces. The committee should consider any matters that are referred to it by a health and safety representative.

Tasmania

The *Workplace Health and Safety Act 1995* and the *Workplace Health and Safety Regulations 1998* provide for both employee safety representatives and committees.

The legislation covers the election of employees' representatives where 10 or more people are employed, and the establishment of health and safety committees where there are more than 20 workers and the majority of them request it.

The functions of safety representatives are to encourage safe work practices, represent the health and safety interests of employees, and bring any hazards to the employer's notice. The representative may inspect the workplace or accompany OHS inspectors, and has the right to access all information reasonably available to the employer relating to hazards at the workplace and the results of any inquiry into accidents, injuries and occupational illness at the workplace.

The number of committee members and the committee's composition are to be agreed between the employer and the majority of employees. Unless agreed otherwise, not less than half of the committee are to be employees elected by their fellow employees. The committee's overall

¶204

function is to facilitate consultation and cooperation on health and safety matters between the employer and employees (sec 28). It makes recommendations regarding training and promotion of health and safety, and the establishment and monitoring of appropriate programs. A person nominated by the committee may inspect the workplace on giving reasonable notice to the employer (sec 29). The employer must, among other things, make information concerning workplace health and safety available to the committee; consult with the committee on proposed changes to the workplace which may affect health and safety; and notify the committee in the event of an accident or dangerous occurrence (sec 31).

Australian Capital Territory

The *Occupational Health and Safety Act 1989* and the *Occupational Health and Safety Regulations 1991* make provision for health and safety representatives and, to a lesser degree, for health and safety committees. The requirements of the Act relating to health and safety representatives only apply to employers who employ more than 10 employees. However, where a principal contractor has substantial control over subcontractors at a construction site, the principal contractor may apply to the Occupational Health and Safety Registrar for a declaration that these requirements apply to that construction site, with the principal contractor becoming the employer (sec 39).

An employer must establish designated work groups at the workplace (sec 37), with each group then selecting one health and safety representative (sec 40). A representative is entitled to time off work, without loss of pay, to attend an approved training program. The employer is liable to pay the fees for the program and any expenses reasonably incurred in attending the program.

Health and safety representatives are entitled to inspect the workplace upon giving reasonable notice to the employer, or when there has been an accident, or when the representative believes that there is a threat of an accident. They are entitled to accompany an inspector on any inspection of the workplace. They may investigate complaints by employees and are entitled to be present at interviews between the employer and employees concerning health and safety. They may have access to information held by the employer regarding health and safety risks at the workplace. Further, where the employer is requested to do so by a representative, he or she must consult with the representative on the implementation of any changes at the workplace which may affect the health and safety of employees.

Health and safety representatives have the power to serve a provisional improvement notice on a person where they believe that person is

Overview of Legislation

contravening the Act, and consultation with the person has not rectified the situation (sec 51). They also have the power to direct that an employee cease work where they believe there is an immediate threat to the health and safety of that employee (sec 56).

The Act makes no reference to the composition of health and safety committees, merely stating their functions. These functions include assisting the employer to develop and implement health and safety measures, facilitating cooperation between the employer and the employees relating to health and safety matters (sec 58), and disseminating appropriate information (in appropriate languages).

Northern Territory

The *Work Health Act 1986* and the *Work Health (Occupational Health and Safety) Regulations 1992* provide for consultation through health and safety committees. If requested by the majority of workers, the employer must establish a health and safety committee, if there are more than 20 workers in the workplace (sec 44A). The committee consists of workers elected by their fellow-workers, and persons appointed by the employer. The total number on the committee is agreed upon by the employer and the workers; half or more must be workers elected by their fellow workers.

The committee facilitates consultation and cooperation between the employer and employees. It must keep itself informed of standards and procedures relating to workplace health and safety; recommend to the employer appropriate programs and procedures; and recommend any necessary training or education relating to workplace health and safety (sec 44C). A person nominated by the committee may inspect the workplace (sec 44D). Employers are required to keep the committee informed about hazards and dangerous occurrences; consult on proposed changes to the workplace; and provide assistance to the committee in the carrying out of its functions.

¶205 Enforcement

Each occupational health and safety Act provides for its enforcement by governmental bodies operating through their inspectorates, with a scale of penalties for non-compliance. Inspectors are required to furnish proof of any failure to comply.

In general, inspectors have the power to ensure compliance through the issuing of improvement notices or prohibition notices. Improvement notices are issued where there is a breach of an Act or the likelihood of such a breach, while prohibition notices are reserved for more serious breaches and generally state that work in the relevant area must cease until the breach is rectified. In some States health and safety representatives are able to play a part in the enforcement of the Act by issuing provisional improvement notices.

Inspectors can initiate prosecutions which, if they lead to convictions, can result in significant fines; and in some jurisdictions they can also issue Infringement Notices (on-the-spot fines). For full details, reference should be made to the legislation referred to at ¶203. Under New South Wales provisions there is also the power for authorised officers of industrial unions to initiate prosecutions.

In practice, prosecutions for the breach of a general duty of care usually occur when an accident, sometimes involving a death, has alerted the inspectorate to an unsafe workplace.

Limitation of enforcement

Although legislation may act as a framework for planning occupational safety and health, it alone is not sufficient. It can require employers to assess manual handling risks but it cannot ensure that the employer will understand or implement this process. What is required is a conscientious commitment by employers, as well as employees and supervisors, to improving health and safety conditions.

Information and education are essential, for larger as well as smaller establishments, where ignorance is often cited as the cause of contravening the law. Education needs to promote a sense of personal involvement and responsibility for safety and health in all those at the workplace. Together with the delegation of specific responsibility to supervising staff and the encouragement of active health and safety committees, it can promote safe work practices.

> Successful health and safety management means going beyond the legislative requirements and adopting a pro-active approach to safety at the workplace.

¶206 Codes and standards

In recent years there has been increased publication of codes of practice dealing with some of the more dangerous industrial areas. These codes are intended to be used in addition to Acts and regulations.

Codes of practice are not legislation and cannot be enforced in the same manner as legislation. However, they do have the backing of the governmental body by whom they are published. A code of practice becomes

Overview of Legislation

effective when approved according to an occupational health and safety statute. An approved code of practice is generally designed to be used in conjunction with the statute and regulations, but a person or company cannot be prosecuted just for failing to comply with it. In proceedings for contravention or failure to comply with the occupational health and safety legislation, failure to observe an approved code of practice may be admissible as evidence. Codes of practice are of particular value to health and safety practitioners because of the detail they provide on safe practice in specific areas.

The occupational health and safety legislation in all States and Territories makes reference to the approval of codes of practice on specific areas.

In addition to the codes published by State bodies, there are National Codes of Practice published by the National Occupational Health and Safety Commission. These codes have no legislative force unless they are adopted into legislation by State agencies, although they can be considered as authoritative guides to good practice in the areas that they cover. The National Codes of Practice have, in some cases, been used as the basis for corresponding State codes. Continuation of this trend will lead to a more uniform national approach to occupational health and safety.

Standards

Like codes of practice, standards are not (of themselves) legislation but are accepted as authoritative guides to good practice. Australian standards have in many instances been incorporated into regulations. Standards tend to deal with specific pieces of equipment or industrial processes.

Australian standards are published by Standards Australia, a non-governmental body (see ¶1002). Each standard is designated by a number, so that standards can be referred to as *AS-XXXX/YEAR* (for example, *AS/NZS 1269 — 1998* is a joint Australian/New Zealand Standard entitled *Occupational Noise Management*). There are thousands of Australian standards, and catalogues can be obtained from Standards Australia. It is worthwhile for employers to obtain copies of Australian standards relevant to their workplace. The National Occupational Health and Safety Commission has taken on the task of developing Australian standards relevant to occupational health and safety.

¶207 Common law liability

Under common law every employer has a "duty of care" to employees and others. This means employers should provide:

(a) reasonably competent staff;

(b) a sufficient number of workers to carry out the work safely;

(c) a place to work that is safe and without risks to health;

(d) proper plant and equipment; and

(e) a safe system or method of work.

An employee has the right to claim damages from the employer for injury or sickness incurred in the course of employment if the injury is due to negligence on the part of the employer or its agents (such as supervisors — see discussion on vicarious liability below). Negligence can occur through acts or omissions. Failure to comply with the duties mentioned above can amount to negligence.

Common law is based upon the decisions laid down by previous cases, which have built up a substantial body of law. Unlike workers compensation it considers the concept of "fault", and amounts awarded for damages under common law settlements can be substantially higher, depending upon degree of fault, extent of pain and suffering, age, disfigurement, and expected future earnings.

Note that common law and workers compensation cancel each other out; that is, an employee cannot obtain settlements under both for the same injury. Any common law settlement is reduced by the amount of any previous workers compensation settlement where both are applied for. Note that in some States the workers compensation legislation abolishes or restricts the right to common law actions. Common law actions by injured employees have been abolished in the Northern Territory and South Australia, are restricted in New South Wales and Western Australia, and are concurrent in Queensland and Tasmania. In Victoria, common law actions were abolished in 1997 and are due to be brought back in May 2000.

To defend itself successfully against such a claim the employer must prove that the measures taken to prevent such injuries, including those taken by other employees acting as the employer's agents, are in accordance with current good practice and that all reasonable care has been taken. As the techniques of health care and accident prevention have improved and become more widely known, the standard of "reasonable care" has risen.

The fact that conditions at a site were not in accordance with statutory requirements and/or relevant Standards can be enough to establish the liability of an employer.

Overview of Legislation

In certain circumstances, contributory negligence on the part of an injured employee can be grounds for reducing the damages claimed (see Contributory negligence below).

The duty of care owed to employees may extend to independent contractors where an entrepreneur engages independent contractors to do work that might easily be done by employees.

Vicarious liability

An employer is responsible for the actions of employees in the course of employment in relation to injuries caused to other employees or third parties.

An act will be within the course of employment if it is:

- authorised by the employer and a lawful act;
- authorised by the employer and an unlawful act if the employer directs an employee to perform duties in a manner which involves the commission of an unlawful act (which includes, for the purpose of the present discussion, breach of contract and illegal acts), the employer will be liable in damages to compensate any third party suffering loss or injury; or
- an unlawful method of performing what is otherwise an authorised act.

In these cases the employer's responsibility extends to the actions of an employee performed within the scope or course of employment.

Contributory negligence

Contributory negligence is a factor at common law, and a percentage figure is determined by the court which will be deducted from the assessed damages in all States except in a particular instance in New South Wales, where it is not a ground for defence when the action is based on a breach of statute.

A comparison is generally made with what a "reasonable" person would have done, and the degree of variance is determined in relation to whether the person was instructed clearly and specifically, together with other relevant factors.

In this regard, if a person knew of an unsafe condition, yet failed to report it or take direct action to correct it, and during the course of events was injured as a result of it, then it is likely that some reduction in damages would occur to allow for the employee's contribution to his/her injury. Each employee is in effect a health and safety officer, required to report or correct unsafe conditions.

¶207

Occupier's liability

Occupier's liability refers to the duty of care owed by an occupier of premises to the world at large. Thus, it extends beyond employees to anyone else who comes to the workplace. For example, a client or a salesperson who falls and is injured because of an unmarked trench on the premises may succeed in a common law action against the occupier.

¶208 Workers compensation

Workers compensation is a statutory obligation placed on all employers. It aims to provide income and rehabilitation for an employee who is injured and is unable to carry on his/her usual work. Workers compensation also provides some lump sum payments in the case of loss or restriction of bodily functions. The purpose of workers compensation legislation is not to apportion blame, and there is no need to prove fault or negligence in order to succeed.

There is workers compensation legislation in each State and Territory. These Acts, and the bodies that administer them, are set out in the table opposite.

Premiums are calculated on the basis of wages and salaries paid and are adjusted from year to year on the claims experience of previous years. The premium rates vary from industry to industry according to the risk involved. For example, the rate for a clerk is only a fraction of that for a tunneller.

The courts have held in many cases that people doing business as independent contractors are to be regarded as employees for the purpose of workers compensation. Accordingly some suppliers of delivery services and some people engaged in the building industry should be regarded as employees when arranging insurance. Advice on this matter should be available from the local State/Territory workers compensation authority, or from insurance companies.

Rehabilitation requirements

Workers compensation legislation is now placing increased emphasis on rehabilitation. Rehabilitation of injured employees reduces the time for which compensation payments must be made, in addition to being beneficial to the employee.

Workers compensation legislation contains requirements that either oblige or encourage employers to rehabilitate their employees. These requirements are noted at ¶718, in the context of the general discussion on rehabilitation.

Overview of Legislation

Workers compensation legislation and administrative bodies

Federal
Safety, Rehabilitation and Compensation Act 1988
Comcare Australia
Safety, Rehabilitation and Compensation Act 1988
Comcare Australia
Seafarers Rehabilitation and Compensation Act 1992
Seacare

New South Wales
Workplace Injury Management and Workers Compensation Act 1998
Workers Compensation Act 1987
WorkCover Authority of New South Wales

Victoria
Accident Compensation Act 1985
Victorian WorkCover Authority

Queensland
Workers' Compensation Act 1990
WorkCover Queensland

South Australia
Workers Rehabilitation and Compensation Act 1986
South Australian WorkCover Corporation

Western Australia
Workers' Compensation and Rehabilitation Act 1981
WorkCover Western Australia

Tasmania
Workers Rehabilitation and Compensation Act 1988
Workplace Standards Authority

Australian Capital Territory
Worker's Compensation Act 1951
ACT WorkCover

Northern Territory
Work Health Act 1986
Work Health

¶208

Chapter 3

Risk Management

What is risk management?	¶301
"Risks" and "hazards"	¶302
Complex causes of accidents	¶303
The risk management process with respect to OHS	¶304
Planning risk management	¶305
Elements of the risk management process	¶306
Factors influencing the success of risk management	¶307

¶301 What is risk management?

The modern approach to protecting workers from workplace hazards is based on the concept of risk management. Any planned and systematic attempt to reduce the risk of work-related injury and disease relies on an understanding of this concept. Risk management is the term applied to a systematic method of assessing and controlling risks associated with any activity, function or process. The term is defined in Standard Australia's *AS/NZS 4360:1999 — Risk Management* to include the organisational cultures, processes and structures that are directed towards effective management of both potential opportunities of exposure to risk and adverse effects of such risks.

The risk management process has been applied in many different industries to control a variety of risks, ranging from professional liability risks to financial risks. However, its most prominent application has been in the Occupational Health & Safety context. In this context it is usually thought of as a three-stage process: firstly risks have to be *identified*, then they have to be *assessed* and *controlled*. In practice, it can be helpful to break down these three stages into smaller steps, as will be explained in more detail in this chapter.

¶302 "Risks" and "hazards"

In a health and safety context, "hazards" are anything with the potential to harm life, health or property and "risks" are the potential outcomes of hazards and the possibility of injury, illness, damage or loss occurring as a result of hazards. The terminology used here when referring to risk management reflects that of *AS/NZS 4360:1999*, as well as much of the more recent legislation in OHS.

Risk management and legal obligations

As the Commonwealth, States and Territories move away from prescriptive legislation towards performance-based provisions, undertaking the risk management process is becoming a principal requirement for all workplaces. Previously, employers were expected to comply with very stringent rules and regulations, which specified exactly what they could and could not do with regard to workplace safety. This was the traditional "minimum standards compliance" approach referred to in Chapter 1. Since the advent of risk management, however, employers have increasingly been given the power (and the duty) to identify their own workplace problems and to find effective solutions to them.

This new approach in legislation places OHS management firmly in the hands of those who are best placed to understand the risks unique to their workplaces — employers and their OHS managers. Employees and their representatives, together with OHS committees, also have important roles to play in this process.

¶303 Complex causes of accidents

In past decades there was a tendency to blame accidents on workers' carelessness or on "freak" combinations of events, but this type of reaction fails to appreciate the many factors that make accidents predictable — such as lack of training, poor maintenance or the absence of safe systems of work. The contribution of factors such as these mean that accidents, as well as injuries and work-related illness, are ultimately controllable. There is rarely, if ever, a single cause of injury or disease at work. Usually many factors are involved that interact in complex ways. The contribution of each factor to the outcome varies considerably.

By following a risk management approach, each individual factor contributing to workplace injury or illness can be identified and its importance as a contributory factor can be assessed. The process then assists in finding appropriate means of controlling those factors that are identified as important in order to prevent work-related injury and illness.

Risk Management

¶304 The risk management process with respect to OHS

The three stages of risk management referred to above (identification, assessment and control) can be achieved through a sixfold process:

- define the scope;
- identify the risks;
- analyse the risks (the first component of risk assessment);
- evaluate the risks (the second component of risk assessment);
- control the risks; and
- monitor and review the process.

Consultation between the employer and employees on each of the six steps is an essential basis of the risk management process.

¶305 Planning risk management

The planning stage of risk management involves setting objectives to achieve progress in occupational health and safety. This entails setting performance standards, which should cover both organisational procedures and the control of specific risks. Information arising from the planning process is therefore essential for effective implementation.

For example, in planning risk management objectives, information may be sought on current industry standards with regard to the control of specific risks in that sector. This information will be extremely useful later in indicating hazards, which should be watched for during workplace risk identification exercises. Equally, statistical information on industry accident rates may be useful in determining the priority of risks during the risk assessment phase.

Elements in successful planning

To plan effectively for risk management, employers should be familiar with:

- the legal duties to employees, contractors, members of the public and others who might be at risk as a result of the organisation's activities;
- relevant industry, hygiene or other health and safety standards;
- the availability of in-house health and safety advice and support;
- access to relevant outside services and organisations that may be able to provide advice and assistance (such as government agencies, training consultants, emergency services and statutory authorities); and
- the principles of risk management.

Contribution to overall management

Safe working methods must be defined before they can be followed. In this context, risk management planning activities can raise managers' awareness of issues such as:

- the need to develop and define safe work procedures;
- prevailing health and safety risks associated with the types of plant, substances and processes existing at the workplace;
- current industry practice with regard to the control of risks;
- OHS issues associated with the introduction of new processes or production methods; and
- new technology available in areas such as plant safety devices, alternative substances to currently used hazardous chemicals, personal protection and work systems (for example, automation and access control).

Prerequisites to implementation

Certain matters should be resolved during the planning process. Prior to implementing a risk management program the employer should:

- identify the need for information on health and safety issues and ensure that it is disseminated as required;
- ensure that effective methods of communication are established and used to promote safety awareness at all levels; and
- ensure that arrangements are in place (and that they are properly used) to achieve effective consultation between management and employees on health and safety issues.

The people asset

An effective risk management program relies on the selection of competent personnel as core participants. The responsibilities of the health and safety manager, OHS committee, OHS representatives, risk management team leaders and others involved in specific tasks within the process must be clearly defined during the planning process. All of these people must be provided with sufficient information and training to undertake their individual roles effectively.

The steps required to achieve objectives should be planned and the plan broken down into tasks that can be allocated to identified individuals or groups. These tasks must be matched to individual competencies and must conform to overall organisational systems and planning.

¶305

For example, an overall objective to reduce strain injuries will require many supervisors and managers in the organisation to understand the individual tasks required of staff. Their abilities and capacities in terms of workload must be matched to the tasks allocated. In addition, they must be provided with sufficient information, training and support to carry out their tasks efficiently.

¶306 Elements of the risk management process

Defining the scope of the process

The first step in the risk management process is to define the scope of the risk management activity, that is, to establish the parameters of the process including the criteria by which risks will be assessed. Therefore, this step defines the strategic and organisational context in which the remainder of the risk management process operates.

This includes:

- defining the external and internal stakeholders and their objectives;
- defining the organisational context — this is the context within which the risk management policy is to be implemented, including what each person's responsibilities are and what resources are required;
- establishing the risk management context including:
  - defining the scope of the specific activity whose risks the process is intended to manage
  - setting an overall time frame for completion of the process
  - identifying the resources required and distributing the responsibilities for conducting the remainder of the process
  - developing the risk evaluation criteria — these may be legal, social, or financial, or may relate to industry best practice
  - planning the structure of the risk management process into logical elements.

The strategic context of the process is defined by:

- the OHS policy; and
- the risk management plan.

OHS policy

An OHS policy is a written document that clearly indicates the organisation's health and safety objectives and the arrangements for achieving those objectives. It should be prepared in consultation with all staff from senior management to the shop floor and must have the endorsement of the CEO. It should also detail responsibilities for health and safety at the various levels throughout the organisation.

Risk management plan

The risk management plan delineates the responsibilities for carrying out each stage of the risk management process, sets a timetable for implementing the process, establishes the criteria for evaluating risks and sets performance indicators to monitor and review the effectiveness of the risk management process.

Identifying risks

This is a critical step in the risk management process. A risk that is not identified will not be controlled. Accordingly, it is crucial that this step is as comprehensive as possible. Ideally, risk identification will be conducted in close consultation with the people performing the activity.

Risk identification involves the systematic investigation of all-potential risks and identifying and recording the hazards that may be causing them. In simple terms, it means identifying all the possible ways in which people may be harmed. To properly undertake risk identification it is important to understand the nature of hazards, the sources of hazards, the forms in which hazards may arise and which hazards have the potential to harm life, health and property.

Hazards may be present in the workplace environment, equipment, substances and work systems. They can be divided into six groups:

- physical hazards;
- chemical hazards;
- ergonomic hazards;
- radiation hazards;
- psychological hazards; and
- biological hazards.

It is necessary to identify all areas in the workplace and work activities where these potential hazards occur and identify the risks associated with them.

¶306

Risk Management

Identifying the range of risks present in the workplace involves activities such as:

- checking records of injuries, accidents and incidents to identify where accidents are most likely to occur;
- referring to resources such as industry standards and codes of practice that may help to identify risks which occur in your industry sector;
- conducting safety audits;
- establishing a regime of daily, weekly, monthly and annual safety checks and inspections; and
- consulting with employees to find out where they think risks may exist.

A systematic process of risk identification could involve:

- developing a list of the tasks, events and processes involved in activities whose risks are to be controlled. Such a list must be comprehensive enough to ensure that no risks are overlooked by the process and at the same time concise enough to ensure that the process is not bogged down with activities that are beyond its scope;
- undertaking a task analysis, which involves breaking down each activity into consecutive tasks and each task into consecutive steps; and
- identifying all negative outcomes arising from or even incidental to each step, as well as all possible causes of the negative outcomes.

More detail on strategies for identifying risks, such as carrying out a survey of the workplace, are given in Chapter 4.

Once all risks have been identified, they must be assessed. This is a two-pronged process of analysis and evaluation. It involves comparing the risks against criteria set in the risk management plan (taking into account legal and industry standards), finding out the best way to control the risk and ranking risks in order to allocate a budget that deals with the most serious risks first.

Analysing risks

Risk analysis is the process of identifying the likelihood and consequences of an event — that is, quantifying the risk — taking into account existing controls. This does not necessarily mean assigning a numerical value to the risk. It can be done using qualitative, quantitative or semi-quantitative processes. Indeed, it may be inappropriate to assign a numerical value to the risk in some circumstances because it inevitably involves putting a value on human life or limb. Furthermore, there is no point in assigning values if

¶306

those values do not translate into meaningful policy prescription. Ultimately, the aim of the exercise is to rank risks in order of priority.

Each risk must be assessed according to its potential to cause harm. The risk analysis must examine both the likelihood of an accident occurring and the potential consequences of the outcome. These criteria are used to rate risks in priority order — as "high", "medium" or "low" risk.

The assessment must take into account the ways in which people's health and safety might be affected and should be analysed considering factors such as the:

- nature of operations carried out;
- substances used or generated;
- way work is organised;
- layout and condition of plant, equipment and the work environment;
- training and knowledge needed to work safely; and
- control measures currently in use.

Evaluating risks

Risk evaluation is the process of comparing quantified levels of risk against established criteria and parameters to determine if the level of risk is acceptable and then to determine priorities for treating/controlling that risk.

Risks can be placed in one of three categories:

- low risk;
- medium risk; and
- high risk.

Low risk — This risk is considered trivial. Accordingly, no further action is necessary.

Medium risk — This risk is considered unlikely or highly unlikely. Adequate controls are in place.

High risk — This is an unacceptable level of risk. Control measures must be developed and implemented immediately.

Controlling risks

Risk control (sometimes referred to as risk treatment) is the process of:

- identifying a range of options for controlling risks;
- assessing their respective efficiency and effectiveness;

Risk Management

- preparing risk control plans; and
- implementing control plans.

Identifying risk control options

Since risk is a function of both likelihood and consequences, we can control risk either by reducing the likelihood of the event or by reducing its consequences.

Risk control must be achieved using a predetermined hierarchy of controls. The primary aim of risk control is to eliminate the risk, the best way of achieving this being to eliminate the hazard. If this is not possible the risk must be minimised using one or more of the other control options from the hierarchy. *The risk control measure selected must be the highest possible option in the hierarchy.*

The hierarchy of controls

1. eliminate the hazard
2. substitute with a lesser hazard
3. modify the work system or process
4. isolate the hazard
5. use engineering controls
6. use back-up controls such as personal protective equipment.

In many cases, it will be necessary to use more than one control method. Back-up controls, such as personal protective equipment, should only be used as a last resort or as a support to other control measures.

Examples of risk control strategies

Elimination

Eliminating the hazard is clearly the most effective method of controlling the risk. This can be done wherever an alternative way of doing the job or task can be devised.

For example, one of the tasks encountered by many plumbers, electricians, fire service installers and data/communications installers consists of drilling into concrete ceilings to attach fixtures such as hooks for suspended ceilings, or brackets to secure water pipes, electricity cables or data and communications cables.

¶306

This task presents the risk of shoulder, neck and back strain due to the prolonged arching backwards of the neck and back, as well as prolonged elevation of the arms when drilling. The weight of the drill is another factor that can add to the risk of pain or injury, as is the need to work on a ladder. A large number of falls in the building/construction industry are associated with working on ladders.

Innovative methods have been devised to eliminate the need to drill into concrete ceilings. One method requires pre-placement of metal anchors before the cement is poured, and another uses extendable tools and improved adhesives to place anchors after the concrete is set. Both of these methods eliminate the need to drill holes into concrete ceilings, and have additional benefits in that the job takes less time to complete, as well as generating less dust and less noise.

Substitution

The replacement of a hazardous process or material with something less dangerous has been widely used in the context of chemicals or other hazardous substances.

For example, asbestos was highly valued in the past due to its remarkable properties — it is light, strong, flexible and extremely resistant to destruction by heat or mechanical forces. As the world has awakened to the health problems caused by asbestos, however, its use has been phased out and substituted with safer materials such as mineral wool, rock wool, glass fibre and cellulose.

The same strategy can be used with hazardous substances such as pesticides, paints and other substances with toxic ingredients. The effort to select the safest material which will do the job has been observed in many types of businesses that use chemicals, such as dry cleaning, photography and horticulture.

Modification of the work system or process

There are many instances in which this strategy can be used to control health and safety risks. For example, in a dusty work environment such as a wood-working workshop or a building site, wet sweeping or wet cleaning methods for dusty work areas can be very effective in reducing levels of dust in the air (and in the lungs of the workers).

Similarly, in a graphic arts workshop, there may be a risk of inhaling vapours when a hazardous substance is being sprayed onto an object. The risk can be much reduced if the substance is painted on with a brush instead of being sprayed on, as fewer fine droplets of the substance will be floating

Risk Management

in the air the workers are breathing. In this way, by changing the process used, the risk is minimised.

Tools can often be modified as well, for example, water spray attachments on tools such as jackpicks, rockbreakers and scrabbling picks used in quarrying or excavation work can do much to suppress the dust generated by the process.

Administrative approaches such as documenting safe work methods, providing training and ensuring sufficient instruction and supervision can also be used to modify the risks of a job or process. For example, entry into confined spaces with potentially contaminated atmospheres has been responsible for many workplace fatalities. Work in confined spaces can be done safely, however, if the safe work practices set out in relevant legislation and standards are rigorously followed, atmospheric testing and monitoring is carried out as appropriate, and workers are thoroughly trained in the procedures.

Isolate the hazard/risk

This strategy involves isolating or enclosing the problematic work process (or machine, or substance), or physically separating the workers from it. For example, soundproof housing can be built around some noisy machinery, or a process such as developing photographs can take place in a completely enclosed machine.

Sometimes it will be the worker who is isolated, eg situated in a soundproof cubicle in a very noisy environment.

Engineering controls

Engineering control of risks can take many forms. Sometimes a process can be carried out by mechanical rather than manual means, eg. unloading a truck may be done with a forklift if the loads are stacked on pallets, rather than having workers manually lift each box off the truck. (This could also be considered as eliminating a manual handling task.)

There are many examples where engineering methods have been used to change the noise-generating components of machinery and equipment, so that the modified machine can do the same task more quietly. (This could also be considered as modifying a work process, demonstrating that these categories of risk control strategies sometimes overlap.)

Back-up controls

The use of *warning signs* is a type of administrative or back-up control. Statutory provision of warning signs is covered by legislation (within

¶306

regulations), but for reasons of standardisation all signs should also comply with various Australian Standards covering size, shape, wording, colour and use. Placarding of chemical storages is covered by a National Occupational Health and Safety Commission guidance note and the Hazchem Codes (see ¶915). Pictorial signs have the advantage of reaching non English-speaking employees (otherwise signs should be multilingual).

Four colours used in signs identify types:

RED — warning

ORANGE — caution

GREEN — safety

BLUE — information

Manufacturers of signs can supply samples and further information. Placement of signs at the workplace should be considered during the health and safety survey, and before commissioning new or renovated plant or buildings.

Protective clothing and equipment

The least effective risk control strategy involves the use of protective clothing or equipment, but in some instances other strategies are not practicable. In the case of certain industrial processes, there are legislative requirements to provide protective equipment.

Protective equipment should only be used as a solution when work hazards cannot adequately be controlled by other means. Protective equipment may also be used as a temporary measure until a risk can be controlled by corrective measures.

Where personal protective equipment is to be used, a list of employees who require protective equipment should be compiled. Items should be fitted and issued personally, signed for and dated.

Types of equipment

The following list contains the most common types of equipment.

Head protection: safety helmets/caps/hats, hoods.

Eye protection: safety spectacles, goggles, face shields.

Hearing protection: ear-muffs, ear-plugs, disposablewool.

Body protection: aprons (leather, cotton, PVC, rubber), safety harnesses.

¶306

Risk Management

Hand protection: gloves (wrist or elbow length chrome leather, PVC, rubber, vinyl-impregnated, cotton interlock, loop pile, stainless steel mesh), various types of barrier creams.

Foot protection: safety shoes/boots (with steel toecap, steel inner sole ankle or knee length).

Respiratory protection: respirators (single or twin cartridge, canister, disposable, airline types — either half or full face coverage), hoods, self-contained breathing apparatus.

Welding protection: goggles, helmets, hand shields, screens, aprons, coats, leggings, spats, gloves.

Statute law requires "approved" types of equipment to be used in specific circumstances. It is the responsibility of the purchaser to check the suitability and compliance of items before purchasing.

All equipment should be in accordance with the Standards set by Standards Australia, with their trademark stamped upon it. This is not only a matter of quality of manufacture. Certain types of equipment must be of the exact type designed to guard against particular hazards. For instance, welding goggles need to be of the exact shade and specification designed for the particular kind of welding carried out. Respirators are of different types corresponding to the gases, fumes or dusts they are designed to protect against.

It is preferable that equipment which must be used regularly, such as goggles or respirators, be issued personally to each employee, instead of being available for common use. Employees will be more willing to wear, and to care for, their own equipment.

Methods of provision vary. The employer may provide the equipment or subsidise its purchase by the employee. Provisions for maintenance of equipment may also vary. Employers may wish to survey the practices of other firms in their industry or locality.

Equipment not in regular use may be kept in stores for issue when required. It should be kept in clean and hygienic condition. Provision must also be made for equipment for visitors or other personnel.

When protective equipment is essential to protect a worker from risk, its use should be compulsory just as use of machine guards is obligatory.

When introducing the compulsory use of equipment, adequate preparation measures are needed, such as consultation with employees' representatives, education, demonstrations and warning signs. Equipment manufacturers will often provide demonstrations of use of the equipment.

¶306

Above all there should be the force of example. Managers and supervisors should be seen to follow the new rules rigidly.

Assessing risk control options

Conduct a cost-benefit analysis of all the control options identified. When selecting control options, take into account the evaluation criteria set at the first stage of the process.

Preparing risk control plans

Detailed plans of implementation of the viable options will include:

- the time-frame for the implementation of the control plan;
- a budget for the plan;
- responsibility for implementing the plan;
- performance measures for evaluating the plan; and
- mechanisms for verifying implementation of the plan.

For example, if a noise problem has been identified, and a plan is being developed to control the risk of noise-induced hearing loss for the workers, it is necessary to first consider the range of control options, such as modification or enclosure of noisy machinery, reducing reverberations with the use of sound-insulating barriers, isolating the workers from the noise and/or requiring the use of hearing protectors (for more details about noise, see ¶1014 following).

Once the most appropriate control measures have been selected, these should be documented in a plan or program that will set out the time-frame (eg "these measures will be implemented within three months"), name the person (or position) responsible for taking action, state what will be done to evaluate the effectiveness of the measures taken and the means of verifying the results (eg "in three months' time further noise measurements will be taken to verify that noise levels have been brought below 85dB(A)").

Implementing risk control plans

Successful implementation of a risk control plan requires a clear delineation of responsibilities and constant monitoring and review.

Reviewing the process

The risk management process is an ongoing one directed to constant improvement. Risks and controls have to be continuously monitored for their change and effectiveness. There may be internal or external factors that affect the likelihood or consequences of an adverse event, thereby augmenting the risk. This is why the risk management process must be regularly reviewed and constantly repeated, and steps must be taken to redress any flaws. This is to ensure that new risks and those overlooked in the original exercise are identified and controlled.

The monitoring and review process involves:

- systematically checking existing risk control measures to assess their effectiveness;
- collecting data on any new risks that have arisen; and
- formulating new control measures.

¶307 Factors influencing the success of risk management

Networking

To be effective, risk management must build on the existing networks within the workplace. The success of the risk management program depends to a large extent on the efficiency of the management structure, the lines of communication and the cooperation of staff and management at all levels. All staff must be given training in how to use the workplace structure to implement risk management.

Fundamental elements of risk management networking and communication include:

- establishing lines of responsibility for risk management throughout the organisation;
- scheduling daily, weekly and monthly planning or review meetings from the shop floor to management levels where risk management is an agenda item;
- conducting meetings of the workplace OHS committee(s);
- undertaking daily, weekly, and monthly walk-through safety inspections;
- ensuring regular liaison between the OHS department (which may include the OHS manager, safety officers, first aiders and a rehabilitation coordinator) and the rest of the workplace;
- ensuring OHS information is distributed widely;
- establishing systems for reporting and recording the existence of hazards, accidents, injuries and near misses (for more detail on incident investigation, reporting and recording see Chapter 6); and

- ensuring effective transfer of relevant OHS information between levels of management, departments and staff.

Management structure and control

Clear allocation and definition of responsibilities help to ensure that:

- personal health and safety objectives are integrated into the overall organisational policy;
- legal duties are clearly defined;
- relevant information from the techniques employed to measure health and safety performance is collected and analysed ;
- individual training needs can be identified according to defined roles and responsibilities;
- relationships between those with different health and safety responsibilities are clearly understood and channels of communication established and used;
- safety awareness and accountability are highlighted and promoted throughout the entire organisation ;
- appropriate criteria can be defined for recruitment and selection of new employees; and
- if the risk assessment process identifies some immediate problems of a sufficient degree of risk, there is a clear understanding of whom the problems should be referred to for action.

Ensuring effective communication

An efficient system of communication is essential for management to retain control of operations. In this way, management can verify that its intentions are understood and followed — otherwise employees may develop their own interpretation of the management agenda. In addition, communication should also involve management listening to the views of employees and OHS experts and incorporating these views into management systems. Effective two-way communication will generate an environment in which all members of the workplace community can contribute to health and safety, thereby helping to make high standards a commonly accepted goal.

In a successful organisation people understand their responsibilities and are competent to fulfil them. Communication of important issues throughout the workplace can be achieved through written material distributed via means such as bulletin boards, newsletters and memos. Meetings and face-to-face discussions are another important way of channelling important information.

¶307

Risk Management

Involvement of all personnel

To achieve optimal success in the risk management process, efforts should be made to involve everyone in the workplace where they can contribute. Commitment to workplace health and safety as a whole improves if the role, progress and results of the risk management process are made available to everyone in the organisation. This can only be achieved through effective communication and networking at all levels in the workplace.

All employees should be consulted and kept informed about:

- schedules of inspections and safety audits;
- upcoming safety, OHS committee and production meetings (including agenda items);
- resolutions of safety, OHS committee and production meetings;
- lost time injury statistics;
- any purchases of safety equipment, plant or substances which will affect safety;
- any changes in the workplace which will impact on OHS issues;
- the hazards and associated risks involved in their individual duties; and
- the control measures implemented to control these risks.

Chapter 4

Planning a Health and Safety Program

Techniques to assist with risk identification	¶401
Some basic questions to ask	¶402
Who should undertake the health and safety survey?	¶403
Identifying specific organisational problems	¶404
The health and safety audit	¶405
Content of a health and safety program	¶406
The effect of job design	¶407
Ergonomics	¶408
Catering for employees with disabilities	¶409
Checklist — the health and safety survey	¶410

¶401 Techniques to assist with risk identification

Methods of identifying hazards and risks to health and safety have already been touched on in the previous chapter, "Risk Management". In any organisation, the methods chosen should reflect the nature of the enterprise and the most effective and feasible way of meeting the objective.

One important technique is to conduct a thorough survey by trained personnel (see also ¶403) to determine existing practices, standards and conditions in each work activity. Employees can also play a part in this survey (see ¶403). The survey can be taken either of the whole organisation at once, or of one section at a time.

The scope of this initial survey has a "one-off" character as distinct from the ongoing activity of health and safety auditing and more routinely undertaken monitoring. (Ongoing auditing of health and safety is discussed in Chapter 8.) Information gathered in this survey can form the basis of an organisation's ongoing control of health and safety.

An initial survey should include consideration and/or measurements of the following:

- lost time and minor injuries (obtained from workers compensation claims, injury report forms and first aid records) also showing cost differences, as the most frequent incidents need not be the most costly, and vice versa;
- sick leave days lost (from human resources/pay office records);
- damage to plant and equipment (from maintenance records);
- workers compensation and insurance arrangements and costs (from the accounts section);
- the background of employees in terms of such aspects as personal disabilities (such as the need for prescription spectacles, the incidence of respiratory ailments), cultural background (whether English-speaking and whether any other languages are spoken) and extent of work experience. In some situations, age distribution may also be relevant (several studies have suggested that the inexperienced, usually younger, worker is most at risk of injury, particularly during the first six months on the job);
- job design (from observation and discussion consider safety, ergonomics, training and work activity layout);
- compliance with legislative requirements;
- types of equipment and raw materials used by the organisation, with emphasis on toxic and flammability levels and other possible health hazards (consider the quality of information available on hazardous substances — Material Safety Data Sheets are discussed at ¶914);
- storage and disposal of waste, including any stockpiling of dangerous/hazardous materials;
- provisions for fire safety (fire alarms, extinguishers, fire-drills, trained employees, areas where fires are likely to start); and
- current risk control measures, and a list of protective clothing and equipment used at the workplace.

All of these sources and records should provide relevant information about current organisational occupational health and safety practices. From this it should be possible to compile an overview of the accident and injury trends and probable areas of risk that require closer attention, supervision and control.

¶401

Planning a Health and Safety Program   63

¶402 Some basic questions to ask

Some basic questions, to form a starting point for an investigation, are suggested below. A more detailed checklist of items to take into account is set out at the end of this chapter at ¶410. Note that each organisation should formulate a range of specific questions in relation to the type of business it is engaged in.

- What kinds of accidents/injuries/damages are occurring?
- How frequent are the various types of accidents/injuries/damages? What are the costs incurred?
- Where are the accidents/injuries/damages occurring?
- Which occupations of employees are related to the accidents/injuries that are occurring?
- Which accidents/injuries/damages cause the greatest time loss?
- What objects and/or substances are most associated with accidents/injuries/damages?
- What factors are associated with accidents/injuries/damages, and is there a recurring item?
- Are there documented safe work method statements, or safe operating procedures, and are these used in practice?
- Is there a system of accountability and are clear guidelines in place in all relevant sections?
- What are the procedures used for accident reporting and investigation? How effective are these?

The information that is collected may reveal weaknesses within the management control system that require immediate attention. So, it may provide a factual foundation on which a health and safety program can be developed to meet specific needs. As the collection of this information may be a considerable task, some thought should be given to a system of storing the information so that it may be easily referred to and updated in future.

As well as employee health and safety it would be wise to consider all other areas in which the organisation has a duty of care, such as product safety, environmental safety and the safety of the general public.

¶402

¶403 Who should undertake the health and safety survey?

If the survey is done in-house, it should be carried out by people with training in safety and/or health matters. Even if the organisation has staff with full-time involvement in these functions, an assessment should be made to establish whether they have the skills to carry out this type of activity. If they do not, training or retraining should be arranged first.

Some companies use outside consultants for some surveys and their own employees for others. This depends on the nature of the survey and the availability of on-site expertise. The combined approach has the benefit of achieving a balance between specialist or expert input and the development of a company's own employees.

Health and safety personnel, both in-house and outside consultants, are considered in Chapter 5.

Organisations can arrange for different types of surveys from outside sources:

- insurance brokers or insurance companies with whom the organisation has a policy;
- private health and safety consultants, including State Divisions of the National Safety Council of Australia (see ¶1203);
- various federal and State government departments (see ¶1202); and
- some safety equipment suppliers.

Role of employees in the survey

Where health and safety representatives and/or committees are established at the workplace, it will be useful to involve them in the survey. This may take the form of consultation in the design of the survey, or involvement in carrying out the survey. As employees, representatives/committee members will be able to provide information on problem areas and common occurrences. Further, involvement of representatives/committee members should lead to a greater level of interest on the part of the other employees.

Each State and Territory makes specific provision for a consultative style of workplace inspection in its occupational health and safety legislation. The relevant provisions in each State/Territory are summarised at ¶204. In general, health and safety representatives (or committee members in New South Wales and the Northern Territory) are able to inspect the workplace on giving reasonable notice to the employer. In some States provision is also made for a representative to accompany an inspector on a tour of the workplace.

Planning a Health and Safety Program

Outside any specific legislative provisions, it is suggested that members of a health and safety committee will need to undertake an inspection of the workplace to familiarise themselves with details of its layout, procedures, problems, hazards and other relevant matters.

¶404 Identifying specific organisational problems

While it is both possible and useful to conduct a health and safety survey along a broad range of general guidelines (for example, as shown at ¶410), many organisations will have their own specific health and safety problems that may require further study and/or outside professional assistance.

Some examples include the following:

- particular types of work and work processes (such as abrasive blasting, spray painting, mining and tunnelling);
- working with dangerous or harmful substances (such as toxic or flammable chemicals), radiation and hazardous dusts (such as asbestos or silica);
- unusual working conditions (such as heat or cold exposure) or animal communicated diseases (such as brucellosis); and
- language or other communication problems (such as mixed ethnic groups, noisy situations or partially deaf workers).

In other cases the initial health and safety survey may alert an organisation to particular problems at the workplace. Either way, the particular problems will require special attention and should be considered in the planning process. A number of the most common safety problem areas are considered in Chapters 9, 10 and 11.

¶405 The health and safety audit

In financial management, there is a projected budget for which capital funding has been allocated. Receipts for expenditure are measured against that known amount to determine whether the organisation is on target for the particular period. A health and safety audit is a similar management tool to measure performance and costs. It is the next stage of development from the checklist system. The latter tells "what to look for", but the audit requires that an item must be converted to *a measurable standard* with its own rating procedure.

A health and safety audit is a systematic and periodic review of the whole OHS management system, including the policy and programs used to promote OHS and prevent workplace accidents/incidents and work-related illness. An OHS audit measures the performance of the whole system to

¶405

determine if the OHS systems are in place and if they are working, that is, if the control strategies are effective.

Because safety auditing is an ongoing activity, it is dealt with in Chapter 8 under the topic, "Evaluating health and safety performance". It is worth mentioning here that the initial health and safety survey can be used to establish standards against which future audits can be measured. Many aspects of occupational safety can be described in quantitative, rather than qualitative, terms. The number of injuries at a workplace and the time lost through injuries are obvious examples. Other examples of quantitative measurements include the levels of contaminants in the air, noise levels and the percentage of time for which any required safety equipment is worn.

¶406 Content of a health and safety program

This section provides an outline of what an organisation needs to implement and monitor in order to achieve a safe and healthy working environment for employees.

It should be noted at this stage that identifying the real cause of an accident can be a complicated process as there may be several coincidental causes creating a chain of causation factors, nona of which in itself would have resulted in an accident. The aspects of accident causation theory are explained in greater detail at ¶603. See *Components of a health and safety program* opposite.

¶407 The effect of job design

It is important to understand the potential effects of job content and job design in order to identify any risks arising from these aspects of work.

Employees may quickly lose interest in a boring, mundane, monotonous job (particularly one over which they have little control or discretion. The likely results of such situations are dissatisfaction with a job and alienation. This in turn may lead to consequences such as lack of attention, tiredness, daydreaming/distraction, lack of care, errors and even sabotage. The adverse effects on safety, costs and productivity are obvious, and they can be compounded by other unsafe or unhealthy aspects of the working conditions.

Planning a Health and Safety Program

Components of a health and safety program

- Involvement and backing of top management, including a system of accountability and health and safety auditing;
- a health and safety policy statement issued by top management, and disseminated to all personnel within the organisation and to contractors;
- health and safety training of employees at induction stage;
- on-the-job health and safety training of employees;
- task analysis to assess the impact of the job on the individual;
- health and safety training of supervisors and management;
- observance of legislative and award provisions;
- a set of workplace health and safety rules, backed by sanctions;
- inspections at regular intervals;
- attention to safety "housekeeping", such as hazard removal, proper storage, tidiness and cleanliness;
- risk identification, assessment and control;
- use of incident/accident investigation techniques;
- accident recording and reporting systems;
- active promotion of health and safety among employees by keeping them informed and by active support of health and safety representative/ committees where these are established;
- provision of adequate first aid and treatment facilities;
- attention to ergonomics and safety and health matters in every workplace and job design and layout, including building design;
- provision of protective clothing and equipment where required, together with training in its use, care and maintenance;
- substitution, where possible, of non-harmful substances for potentially harmful ones;
- removal of potential contaminants, such as dusts and vapours, at the source;
- attention to the needs of employees with disabilities (for example, work layout and facilities);
- monitoring of potentially hazardous environments by an industrial hygienist to control atmospheric levels;
- preparation for emergencies, including fire, bomb and catastrophe;
- investigation of "near misses", as well as accidents that involve injury or property damage;
- provision for worker involvement and participation with a feedback loop to senior management. There should also be feedback to employees on changes and improvements so that there is an awareness of achievements as well as problems;
- a program of health care beyond basic employment requirements (for example, fitness programs); and
- counselling facilities for health problems, such as stress or alcohol/drug dependence.

¶407

Many factors may contribute to unsatisfactory job design and content, including:

- lack of control over the job, a particular problem with assembly-line situations and where technological enhancement results in "de-skilling" and/or where the operator has to meet the pace set by the new technology;
- inability to use potential and initiative;
- lack of "relevance" of work — the worker is unable to see the point of the work or its end product;
- lack of variety in work tasks, a problem that can often be overcome by introducing job rotation;
- a mismatch between job needs and opportunities with individual ability, skills and aspirations;
- social isolation (either spatial or ostracism); and
- lack of feedback and recognition on the job.

Basically, these add up to a conflict between the needs of the employee and the needs of the organisation and may result simply from an unimaginative approach, or through management being unaware of potential problems. Consultation with employees through health and safety representatives or health and safety workplace committees on matters such as the implementation of changes may obviate many such problems.

When designing jobs and allocating employees to them, attempts should be made to align individual and job needs as much as possible. Popular techniques include job enlargement and enrichment schemes, attitude surveys, forms of employee participation in decision making, quality circles and the use of autonomous or semi-autonomous work groups. Remember that the more interested and satisfied an employee is with the job, the more chance there is that the job will be performed efficiently and *safely*.

¶408 Ergonomics

Ergonomics was defined at ¶109. As indicated by the definition, it covers a wide range of workplace aspects — location, comfort, layout, ease of usage and colours. In any health and safety survey the layout of each workplace should be considered. The employees concerned can usually provide information on any inadequacies of design or layout.

As ergonomics is a very complex science, it is advisable to seek expert advice when purchasing or installing new equipment, instituting new work processes or changing the workplace layout.

Planning a Health and Safety Program 69

Personal observation of and interviews with employees prior to introducing change or purchasing new equipment provides useful information to highlight deficiencies — a health and safety representative and/or committee provides a mechanism for consultation for such purposes.

¶409 Catering for employees with disabilities

An important part of a health and safety survey is to ensure that the workplace is safe and convenient for any current physically disabled employees (or suitable for future ones). This refers to a wide range of disabilities and is a consideration that may become more relevant with an increasing emphasis on the rehabilitation of employees. As what will be required varies according to the circumstances of each individual, it is only possible to provide general guidelines on what needs to be considered.

Important information to be aware of includes State building regulations and Australian Standard 1428 *Code of Practice for Design Rules for Access by the Disabled*.

Items to be considered include:

- building layout — parking, ramp entrances, doors, lifts, washrooms, level entrances and adjacent floor levels;
- work layout — desks, shelves, drawers, chair adjustment, workbenches; and
- work equipment — a wide range of equipment specially designed for disabled people (such as those with sight disabilities or limb amputations) is available.

Advice is available from manufacturers of equipment and organisations involved with disabled people. It can also be helpful to consult the employee, who is in a very good position to know what he/she can and cannot do, and what assistance is needed.

¶410 Checklist — the health and safety survey

The aim of this checklist is to demonstrate the aspects of the workplace that should be examined when surveying an organisation in order to plan an occupational health and safety program. Note, however, that some aspects are now specifically referred to in legislation (for example, the right of employee representatives to inspect the workplace). The checklist provides guidelines only and, when designing a checklist for a particular workplace, individual circumstances and relevant legislation must be taken into account.

To facilitate its use, the checklist is set out in questionnaire form.

Checklist — the health and safety survey

A. Management responsibility and staffing

1. Is there a written occupational health and safety policy (see ¶702) signed by the most senior person? Is it publicised throughout the organisation? Is it disseminated to contractors/suppliers? What are the policy contents? How often is the policy reviewed?

2. Top management — list the extent to which they participate in the health and safety function and assist in its administration. How do they set a good personal example?

3. Are persons engaged in the following categories:

    (a) occupational health and safety officers/managers — full/part-time;

    (b) first aid attendants;

    (c) occupational health nurses — full/part-time;

    (d) occupational physicians/company doctors — full/part-time or on a contract basis; and/or

    (e) workers' health and safety représentatives?

    To whom do these persons report?

    Are their duties, responsibilities and authority clearly set out?

    Are the locations of these people suitable?

    Do they have adequate equipment, time and facilities?

    What training is provided for them?

    How effective is that training?

    Where legislation exists, is it being complied with?

4. Are there health and safety committees?

    If so, how are they staffed (management only or joint employee/trade union/management)?

    What matters are included/excluded from their attention?

    To whom do they report? How much influence do they have?

    How often do they meet?

    What has been achieved by them?

    What training is provided for committee members?

    How effective is that training?

    Where legislation exists, are the relevant provisions complied with?

¶410

Planning a Health and Safety Program

If there is no committee in operation, what forms of communication on health and safety matters exist between employees and management?

5. How often (and to what extent) are health and safety discussed at board and management meetings?
6. Are there written operating rules or procedures on health and safety?

    If so, how are they communicated to employees?

    How often are they reviewed?

    If some employees are not proficient in English, are written rules or procedures also issued in other languages or is the information conveyed by non-written means, as appropriate?
7. How and to what extent is responsibility for health and safety delegated to:

    (a) line managers;

    (b) supervisors; or

    (c) rank-and-file employees?

    Are there any problems caused by the approach(es) taken?

B. Accountability for health and safety

1. How does management hold all levels of managers and supervisors accountable for risk management?
2. Is there a system of costing accidents/work-related illness and compensation claims (for example, by section or division)?
3. Do performance appraisals of supervisors include discussion on their section's health and safety performance?
4. Are supervisors responsible for regular equipment, facilities and machinery inspections, accident investigations, and health and safety training sessions? What means are there of ensuring that supervisors carry out these duties?
5. What type of system operates to encourage suggestions and comments from staff and operatives (for example, a representative from each section on a safety committee)?

C. Inspecting the workplace

1. What inspections are made of the workplace layout, conditions, machinery/equipment and facilities? How is compliance with health and safety rules and procedures enforced?

¶410

2. What systems are used for risk identification:
    (a) direct observation;
    (b) critical incidents;
    (c) sampling techniques;
    (d) statistical reports;
    (e) interviews/questionnaires; and/or
    (f) others?

    Is there a need for further training in these techniques?

    How is each job broken down into component elements and movements to determine the location of any risks? How useful are the records kept from this analysis?

3. Are risk assessment and control records kept and, if so, have they been appropriately followed up and monitored?
4. Who is responsible for the inspections? To whom are the results reported?
5. What type of follow-up action is taken? By whom?
6. What procedure is followed to ensure safety of new plant, processes, operations or materials used? Is health and safety considered when purchase or installation is contemplated? (The success of initiatives here will often depend on gaining the cooperation of suppliers. It may be useful for organisations engaged in similar areas to formulate common health and safety oriented guidelines to assist suppliers.)
7. When corrective action is found to be necessary, how is it initiated and carried out? Is there a written follow-up system?
8. The list below represents items or conditions that may be found in a workplace. Obtain details of the type of inspection made and its frequency.

Item	Frequency of inspection	Date last inspected	Comments
Equipment maintenance			
Noise levels			
Lighting			
Ventilation			
Temperature			
Humidity			

Planning a Health and Safety Program

Item	Frequency of inspection	Date last inspected	Comments
Extraction systems			
Air-conditioning			
Fumes			
Cleanliness/hygiene			
Stairways			
Passageways			
Floor coverings			
Lifts (incl marking maximum load)			
Scaffolding			
Hoists and cranes (incl marking maximum load)			
Ladders			
Ropes			
Chains, hooks and slings			
Forklift trucks			
Other in-plant vehicles			
Boilers (incl certification certificates)			
Ovens/furnaces			
Machinery guards			
Power presses			
Lathes			
Portable grinders			
Pedestal grinders			
Other woodworking machinery (circular saws, routers, joiners)			
Emergency isolation of machines			
Power conveyors			

¶410

Item	Frequency of inspection	Date last inspected	Comments
X-ray equipment			
Drums and belts			
Control of hazardous substances			
Power tools			
Hand tools			
Qualifications of operators (boilers and pressure vessels, crane, forklift)			
Compressed air (air lines, vessels, valves — incl certificates)			
Electrical equipment (insulation, wires, switching, earthing)			
Emergency and operating procedures (incl display)			
Vehicle safety procedures			
Checks on vehicle users, licence inspections			
Washrooms/facilities			
First aid kits and contents			
First aid training			
Warning signs (incl multilingual)			
Precautions for confined spaces			
Protection from dangerous materials and substances			
Issue of protective clothing and equipment			
Maintenance of protective clothing and equipment			

¶410

Planning a Health and Safety Program

Item	Frequency of inspection	Date last inspected	Comments
Materials handling, incl: (a) lifting and gripping (b) storage (c) internal transport routes			
Floor loads			
Pits and excavations			
Radiation			
Vibration from equipment			
Provisions for people with disabilities			
Non-smoking areas			
Office furniture			
Any other items covered by legislation/awards or peculiar to this workplace			

D. Employee selection

1. Check the relevance of recruitment application forms and recruitment interviews to health and safety issues.

2. How are references and previous employment history checked?

3. Is a pre-placement medical examination given? (Note the distinction between pre-placement and pre-employment — in the former case, the employee is to be hired and a suitable position is being sought.)

4. Does the work involve any activities for which subsequent medical examinations are required for health surveillance purposes, and are subsequent medical examinations of employees arranged? Or are employees encouraged (not compulsorily) to arrange them themselves?

5. How is the information from medical examinations recorded?

6. Are any psychological or motor ability tests given at the recruitment stage?

7. Are any physical requirements of the job contained in the job specification? How are they taken into account when recruiting?

¶410

E. Training and induction (see generally ¶706 et seq)

1. Do new employees receive health and safety instruction or training during induction? What form does this training take?
2. When a new employee is being trained on the job:
    (a) who does the training;
    (b) how is it done;
    (c) are there written job instructions, work method statements, safe operating procedures, or some form of training manual used; and
    (d) do the instructions include health and safety aspects?
3. Does the organisation have a written set of health and safety rules?
4. Is any training given to current employees who are transferred to different jobs?
5. Is use made of outside training courses (such as those conducted by government departments or technical colleges)? Are these courses evaluated?
6. Is there any provision for extra training of non-English-speaking employees (such as the issue of instructions in multilingual form, or the conducting of English-speaking classes wholly or partially in work time)?
7. What methods for training supervisors are used? Is health and safety instruction included in the training? Is the training evaluated?

F. Employee motivation

1. Which of the following approaches, if any, are used to motivate employees to observe health and safety requirements? How have they been evaluated to assess their contribution towards improved health and safety?

Item	Details	Evaluation
Group meetings		
Published material		
Films/audio-visual		
Posters		
Warning signs		

Planning a Health and Safety Program

Item	Details	Evaluation
Lectures by outside experts/specialists/authorities		
Publication of statistics		

2. Have any special campaigns been adopted in the past (such as eye protection campaigns or "family nights")? How successful were these?
3. Are such initiatives of a "one-off" kind or an integral part of an ongoing program that provides regular follow-up and reinforcement?
4. How can various jobs be made more interesting for employees (and therefore more satisfying and health and safety enhancing) by removing or reducing their monotonous or repetitive aspects?

G. Protective clothing and equipment

1. List the areas and work processes identified where protective equipment is required. Are there any other areas?
2. In areas where protective equipment is used, can the risk be controlled by alternative methods, higher in the hierarchy of risk control, to make protective equipment unnecessary?
3. What approved types of equipment are provided? Are they appropriate? Is the currently used equipment now superseded?
4. What are the award and/or legislative provisions regarding protective equipment?
5. Is the equipment comfortable and convenient to use?
6. Is the equipment maintained in good condition? What arrangements are made for maintenance (such as laundering of clothing or replacement of cartridges in respirators)?
7. What arrangements are there for the supply and issue of equipment?
8. Is the equipment actually being used? What means of enforcement and supervision are available?
9. Is the equipment properly fitted for each individual (for example, are beards interfering with the proper fit of facemasks)?
10. Do employees fully appreciate the function of a particular safety device? (For example, the effectiveness of hearing protectors is, in part, a function of the length of time these are worn when exposed to noise. Damage to hearing can occur with exposure to levels well below that which might be felt as "uncomfortable".)

¶410

H. Safety and health research

The quality of information available dictates the quality of a health and safety program based on that information.

1. Are the services of an outside professional organisation used? Is it contacted often? How useful are the services and advice provided?
2. List the safety and health journals and publications to which the organisation subscribes. Are they being read, and is the information being used? Should the circulation list within the organisation be widened?
3. What are the main sources of the organisation's information on health and safety matters?
4. Who carries out the research on health and safety matters (if any) on the organisation's behalf?
5. Is useful information being obtained from employers' associations, insurance companies, government departments or other sources?

(A regularly scheduled occupational health and safety committee meeting with both employer and employee representatives provides an opportunity for all interested parties to receive the same information at the same time in the same way. This is known as "information parity" and the experience of other countries which have operated similar schemes for some time suggests that this is valuable, if not critical, to the implementation of a successful health and safety program.)

I. Fire safety (see generally ¶921 et seq)

1. How (and how effectively) is employee instruction carried out?
2. How often are fire-drills held?
3. Are fire wardens and/or fire-fighting teams appointed?
4. Are fire exits clearly marked (and not blocked or locked)?
5. Are fire alarms installed? How often are they serviced or checked?
6. Are fire extinguishers accessible, with usage instructions clearly set out? How often are they checked?
7. Are precautions taken to reduce the risk of fire outbreaks (for example, due to inflammable materials not properly stored)?
8. Does the organisation have an emergency evacuation plan?
9. Are all legislative requirements complied with?
10. How often are inspections held?

¶410

Planning a Health and Safety Program

J. Accident investigation (see generally Chapter 6)

1. What are the most common types of accidents?
2. What are the most common types of injuries?
3. What conditions and circumstances determine which accidents will be investigated, if some are not?
4. Who carries out these investigations?
5. How are the investigations carried out? Are there set procedures?
6. What types of accident investigation reports are prepared? To whom are they submitted?
7. What follow-up action is (or can be) taken? Who takes the action? Has it been effective in preventing a recurrence of the accident?
8. Have any specific risk control programs been established? How successful are/were they?
9. Are the current report or investigation forms adequate?

K. Health and safety information, accident records and analysis

1. What records and statistics are kept (such as accident frequency and incidence rates, compensation claims, or accident reports)? By whom are they kept?
2. Who uses and/or has access to these records?
3. How often are records analysed (daily, monthly or yearly)?
4. Are risk assessment and control records kept? In what form are they kept? What forms of follow-up on them occur?
5. Is a cost analysis technique applied to the records?
6. Are time comparisons kept to provide an indication of safety/accident trends within the organisation?
7. Is there any follow-up study resulting from the statistics kept? By whom is this done?
8. Are accident report books kept in all work sections, in an appropriate form?
9. Has the viability of computerising such records been investigated? What computer resources are available?

L. Medical programs and facilities (see Chapter 5 for more detail)

1. Which of the following staff are employed (numbers):
    (a) first aid attendants (in each work section);
    (b) occupational health nurses;

¶410

(c) occupational physicians; and/or

(d) occupational hygienists?

2. What makes these people appropriately qualified?
3. How often are first aid training courses held? Is this sufficient?
4. Who is responsible for first aid supplies and facilities? Does this arrangement work effectively?
5. What is the procedure for obtaining first aid assistance?
6. Are first aid facilities suitably located?
7. Are medical treatment facilities adequate and kept in good order?
8. What emergency facilities are available if the normal first aid attendant is unavailable?
9. What facilities are available for transportation of the injured to a hospital?
10. Is a directory of qualified physicians, hospitals, ambulances available?
11. Are any health education programs carried out by the organisation? How have these been evaluated?
12. Are there any disaster or "catastrophe" plans prepared by the organisation?
13. What leisure facilities are available (such as sporting, gymnasium or recreation areas)? Are these promoted by the organisation? How often are they used?
14. Are counselling and/or referral facilities available to employees with health problems such as stress or alcohol/drug addiction? (See generally ¶1106 et seq.)
15. Does the organisation tend to promote a "healthy lifestyle" for its employees? In what ways? (See also 13 above.)
16. Are there any particular health risks/problems at the workplace, such as harmful substances/work processes? How well do employees working with these understand the hazard(s)? What action can be taken to minimise the effects of these?
17. Are there warning signs of problems, such as "disease patterns" (for example, a number of employees suffering from headaches, nausea, rashes, eye irritations or coughing)? Are these investigated?
18. Where health hazards exist, are there adequate means of warning employees?

¶410

Planning a Health and Safety Program

M. Miscellaneous

Any other aspects of the workplace not mentioned above that require action or attention.

Item	Comments

¶410

Chapter 5

Staffing the Health and Safety Function

Introduction .... ¶501
Who is responsible for health and safety? .... ¶502
Responsibilities of top management .... ¶503
Responsibilities of line managers and supervisors .... ¶504
The role of the human resources department .... ¶505
The health and safety officer or manager .... ¶506
Health and safety representatives .... ¶507
Workplace health and safety committees .... ¶508
The first aid officer .... ¶509
Occupational health and safety professionals
When to engage an expert .... ¶510
The occupational health nurse .... ¶511
The occupational physician or company doctor .... ¶512
The occupational hygienist .... ¶513
Other occupational health and safety consultants .... ¶514
Establishing a group health service .... ¶515

¶501 Introduction

The purpose of this chapter is to describe the roles and responsibilities of the various types of staff involved in an organisation's safety and health functions.

The usage of these types of staff will depend on the organisation's own circumstances — its size, dispersion of employees, type of work, control systems for hazardous processes, amount of dangerous materials in use, incidence of shiftwork, working conditions, proximity to outside health and medical care, and other individual aspects relevant to its style of leadership and organisational structure.

Whatever staffing format is established, there will be a danger of overlapping of authority or dual subordination (for example, between first aid officers, nurses, doctors and health and safety officers). It will be useful if clear job descriptions can be prepared and issued to these groups.

All employees at the workplace should be made aware of the various safety and health staff and the functions of each. This information should be included as part of induction training.

¶502 Who is responsible for health and safety?

The final responsibility for occupational health and safety lies with the person who is ultimately in charge of the workers, such as a CEO, Managing Director, Chairman of a Board or Permanent Head.

Just as authority is delegated from this top position down through the organisational structure to ensure that the objectives of the organisation are efficiently fulfilled (in areas such as finance, personnel, purchasing, distribution, sales and production), so responsibilities for occupational health and safety need to be allocated and persons held accountable.

The authority delegated to carry out health and safety responsibilities must operate in the same way as other management functions. It begins at the top level with an approved written policy statement, procedures, rules and instructions that, once issued, must have some type of performance assessment system to measure compliance and personal accountability.

To carry out the health and safety function effectively, the degree of authority delegated must equal the amount of responsibility given. For example, those who are accountable for the health and safety of workers in a particular area must have the authority to redesign work processes in that area. While this is relatively straightforward for organisations that employ workers on their own premises, it is more difficult when the employer is a labour hire company whose workers are hired out to other organisations.

The courts have held that in this case the employer, that is, the labour hire company, still shares responsibility for the safety of workers who are employed at other organisations. In this situation the labour hire company as well as the "host employer" are both responsible for the workers' health and safety.

Many organisations evaluate health and safety performance, such as accident prevention results, as one aspect of performance when considering possible promotion opportunities. Frequently financial loss due to accidents, injuries and damage may equal or exceed the organisation's profit for the same period. Therefore it is advisable that these losses and results be included in all relevant reports, including annual reports to shareholders.

¶503 Responsibilities of top management

Top management will be responsible for an organisation's strategic planning in relation to occupational health and safety. This will mean determining the resources (finances, staff and access to consultants) that are to be devoted to health and safety. What constitutes adequate resources may be contentious, with employees and management expressing conflicting views. Senior management will be required to make the final decision and must take responsibility for an organisation's health and safety performance.

Other top management functions relevant to health and safety include:

- determining which safety programs are to receive priority;
- monitoring the outcome of an organisation's health and safety programs;
- ensuring compliance with legislation, awards and standards;
- endorsing the formulation of appropriate rules, procedures and methods for the workplace;
- ongoing and effective dissemination of OHS information and promotion of health and safety awareness in the workplace; and
- commitment by personal example.

¶504 Responsibilities of line managers and supervisors

Line managers and supervisors must accept as an integral part of their duties the functional responsibility for implementing and administering health and safety procedures at the workplace. Staff specialists are there to assist and advise them, not to carry out their work for them. Line managers and supervisors, however, must have the authority to match their responsibility in order to act effectively.

Supervisors/managers will have overall functional management responsibility over their sections, and the basic functions of planning, organising (including staffing), leading and controlling apply to the safety issue. Responsibilities include the following:

1. overall supervision of employees to ensure the health and safety of the worker, the public and the consumer;
2. hazard identification and assessment and control of risks;
3. implementation of particular safety programs;

4. on-the-job training;

5. efforts to motivate employees to comply with safe work practices, including specific directives when giving orders;

6. control of plant and mechanical/electrical equipment;

7. accident investigation and correct reporting;

8. issue of ensuring correct use and maintenance of appropriate personal protective clothing and equipment;

9. submit reports and recommendations/suggestions about hazard controls, workplace procedures, etc, to more senior management, when these issues are outside their scope of authority; and

10. decision-making regarding job design, workplace layout and (possibly) recruitment, where they have this authority.

Where the organisation has full-time health and safety staff (see ¶506), this staff will provide a support or back-up role to managers and supervisors, but will not have the responsibility of the latter transferred to them.

¶505 The role of the human resources department

The role of a human resources department within an organisation is one of a service/advisory/strategic function, not a line management function. Frequently, the health and safety function is administered by the human resources department, where it is often one of the "extra duties" added to that function. Where this is the case, a specific time allocation must be given to enable those duties to be carried out. Otherwise, this function will be given less time due to other pressures until it becomes a function in name only. Furthermore, it should be clearly understood that the human resources department's activities in administering the health and safety function do not lessen the responsibility of line managers or supervisors.

Where the human resources department is given the task of administering or resourcing the health and safety function, some or all of the duties set out at ¶506 should be performed by that department. In this situation it will be worthwhile to build up a list of outside contacts and sources of technical and health and safety information, as it is likely that this type of assistance will be required.

The human resources department may participate in a workplace health and safety committee.

It is also appropriate for this department to be aware of the functions of occupational health and safety staff and of training available in this area. The department may be required to coordinate training for staff.

Staffing the Health and Safety Function

¶506 The health and safety officer or manager

This section outlines the factors involved where the organisation employs full-time staff within the health and safety function.

(Note that the Queensland legislation requires the appointment of a "health and safety officer" at each workplace with 30 or more workers (see ¶203). The legislation appears to envisage a senior management position, that would include some or all of the functions discussed in this paragraph.)

As health and safety has taken on a greater prominence in workplaces, the designation "health and safety officer (or manager)" is tending to replace "safety officer" and the narrower focus that that title implies.

As can be seen from the list of the person's likely duties below, the job is mainly an administrative and advisory one (that is, a staff position). The officer/manager should report directly to top management, who take action after considering the advice received. At the same time, it is necessary for the officer/manager to possess sufficient authority to take any action required in a possible emergency. For example, if the supervisor was not readily available, and a dangerous situation arose in which an accident could occur, action must be taken to prevent injury and damage. Normally, however, the position will have no formal authority over line managers or supervisors. There is greater emphasis in organisations today on all employees sharing responsibility for OHS.

The following list has been checked against a job description for the position of occupational health and safety officer with a Commonwealth Government Department.

Duties of a health and safety officer/manager

1. Conducts inspections personally or in company with executives and supervising officers, specialist consultants, or health and safety representatives/committee members to ensure the observance of health and safety standards and for the purpose of discovering unsafe or unsatisfactory conditions and practices before personal injury occurs. The health and safety officer/manager should be able to identify and qualify/quantify risks of injury or disease from occupational hazards and determine possible control measures.
2. Reports any unsafe and unsatisfactory conditions, procedures or operations to the supervisor or executive in charge (note that authority to order cessation of work in a dangerous situation must receive serious consideration, particularly where there is legislative provision for the cessation of work deemed to be unsafe (see ¶203)).
3. Acts as adviser to executive and supervisory staff and employees in all matters concerning prevention of accidents, injury, hazard, disease and the promotion of health and safety. Monitors the organisation's overall safety performance in order to report this to senior management.

Continued over

4. Receives and reviews all accident and injury reports or reports of any potential hazard or "near misses". All such reports should reach the health and safety officer as quickly as possible. Prepares reports and recommendations to management and/or safety committees/representatives.
5. Investigates selected accidents, hazards or "near misses", as distinct from the obligatory investigation made by supervisory staff, and recommends the appropriate action to prevent a recurrence.
6. Maintains an injury record system generally in accordance with relevant standards.
7. Attends all meetings of health and safety committees, and staff meetings or conferences when health and safety matters are discussed or considered.
8. Organises health and safety training of staff and employees in conjunction with executives and supervisors. Institutes health and safety promotion campaigns to create and maintain an interest in health and safety at all levels. Conducts or arranges health and safety induction courses for new employees.
9. The occupational health and safety officer/manager should be able to identify necessary resources and be aware of the availability of specialist support advice. Where medical or nursing staff are not immediately available on the premises, ensures that first aid facilities are adequate and maintained in satisfactory condition. Also makes arrangements with ambulance, casualty and medical services to be available as required.
10. Monitors and informs — on legislation and award provisions relevant to health and safety. Advises of changes.
11 Reviews, monitors and updates health and safety manuals, rules, procedures, etc.
12. Encourages compliance with safe work practices through distribution of training resources such as literature, posters, warning signs, videos and films, etc. This material can be either general (such as an induction handbook) or on specific topics (such as handling harmful chemicals, manual handling, etc).
13. Liaises with the organisation's insurance company or broker. This person may be involved in liaison in a workplace rehabilitation program (assisting in evaluating the effectiveness of programs, etc).
14. Maintains contact with government departments and authorities to determine changes in legislation and any incidents that may be applicable to their workplace.
15. Stays abreast of health and safety techniques within the particular industry.
16. Carries out or organises research into safety and health matters as required by the organisation.
17. Answers enquiries from management, supervisors and employees.
18. Evaluates health and safety products and systems, and advises management on their suitability and application to the organisation.

¶506

Staffing the Health and Safety Function

The health and safety professional should have the training to be able to recognise most of the types of hazards that exist, and be able to learn by both studying the workplace and consulting with other professionals. Good liaison between the health and safety officer and supervisors is essential as the health and safety officer could lack the technical background of the supervisor.

Qualifications of health and safety officer/manager

When drawing up a job specification for a health and safety officer, either full- or part-time, the following qualities are desirable:

- knowledge of hazard identification, risk assessment and control;
- thorough knowledge of relevant legislation, codes of practice and standards applicable to the workplace, including any regulations covering buildings, fire protection, electricity, vehicles, lifting appliances, certification requirements, harmful substances, etc;
- a capacity to understand the nature of the organisation's activities;
- other technical knowledge, such as storage and control of dangerous goods and hazardous substances, fire prevention and security;
- skills in report writing and verbal communication;
- the qualities of tact, patience, and the capacity for accurate, efficient handling of detail;
- the ability to undertake instruction and training;
- the ability to liaise with the workplace health and safety representatives and committee members;
- knowledge of current developments in disease and accident prevention methods and thinking, both in Australia and overseas;
- understanding accident causation and the techniques used in reporting, investigating and recording accidents and injuries, and accident statistics;
- enthusiasm for health and safety work;
- adaptability and willingness to undertake further training in health and safety, and to engage in outside activities such as part-time health and safety training and membership of appropriate professional organisations;
- a good knowledge of the firm's policies and procedures;
- the ability to distinguish between the advisory function performed as health and safety officer and any other executive or technical functions he/she may have; and

¶506

- the ability to influence personnel, enabling change to be implemented.

The health and safety professional should be given the opportunity to acquire this knowledge and these skills and abilities. Advice on courses available should be sought from the government departments and organisations listed in Chapter 10 as well as university departments and TAFE colleges.

¶507 Health and safety representatives

Health and safety representatives are the most common agents in promoting employee participation in safety at the workplace. Representatives allow employees to express safety concerns to management, and provide management with information which they may not otherwise obtain on potential safety problems at the workplace. Active health and safety representatives will be able to generate an interest in safety throughout the area that they represent.

Health and safety representatives are required by legislation in all States and Territories, except New South Wales and the Northern Territory (where they have health and safety committees (see ¶508)). Generally the legislation sets out the method of selecting a representative, as well as the powers and functions of the representative. This legislation is summarised State by State at ¶204.

Selection of representatives

Generally a health and safety representative acts for one particular group of employees. These groups will often be the various departments or geographical areas within an organisation, but they may also be chosen along other criteria such as shiftworking arrangements or commonality of the hazards faced by particular employees. In smaller organisations it may be sufficient to have only one health and safety representative.

To ensure employees' confidence in a representative, and to guarantee that the representative remains accountable to the group, health and safety representatives should be elected by those they will represent (see ¶204).

An appointment for a period of at least two years is desirable, because of the training necessary and the period required for representatives to familiarise themselves with the role. However, there is also some advantage in a rotation of employees in the position, because of the experience and awareness that it creates in the employee concerned.

Staffing the Health and Safety Function

Involvement of unions/union delegates

An issue that arises in relation to health and safety representatives is whether work groups should be determined along the lines of union membership and whether union delegates should become health and safety representatives. Clearly the health and safety of employees is a major union concern; however, health and safety issues should not become mixed up with more general employer/employee matters, such as dangerous work allowances.

In workplaces that have poor industrial relations records, it may be best not to determine work groups along union lines. To do so could encourage safety issues to be thought of as an extension of industrial relations issues, and may lead to an adversarial approach in relation to the safety issues that do arise.

There is no overwhelming argument either in favour of or against having union delegates as health and safety representatives. Having both positions performed by the one person may lead to the merging of safety and general industrial relations issues that was discussed above. It can also be argued that where one person occupies both positions, he/she will not be able to give sufficient time and energy to either. On the other hand, a good union delegate will already have the necessary negotiation and consultation skills required to succeed as a representative. In reality, a popular and effective union delegate is likely to be elected as a representative (if he/she wishes to take on the additional workload), whilst a poorly performing delegate will not be.

Functions and powers

The crucial functions of a health and safety representative are to convey employee safety concerns to management and to act as a representative of the employees in relation to safety issues. This function's basic concern is to improve health and safety at the workplace.

In order to carry out the function of health and safety representative effectively, the following powers are regarded as essential (in each State, these powers are prescribed to a greater or lesser degree (see ¶204)):

1. *The right to know about health and safety at the workplace.* This means the training of the representative, that is discussed below. It also means access to information that the employer has relating to potential dangers at the workplace. The representative should be allowed to keep up to date with safety developments in the relevant area (courses, subscriptions to journals, tours of similar workplaces).

2. *The right to inspect the workplace.* It is important that representatives conduct regular inspections of the area they represent. A checklist may be useful. The right to investigate an accident or dangerous occurrence is also usually prescribed by the legislation.

3. *The right to participate in health and safety activities.* Representatives should have input into the safety policies adopted at the workplace. They should be informed of any new equipment or processes that are being considered. The legislation may also provide a right to be present at any interview between the employer and an employee concerning a safety matter.

In some States health and safety representatives are given the power to direct employees to cease work where the workplace is considered to be unsafe, and the right to serve the employer with a provisional improvement notice where the representative considers that there is a breach of legislative requirements (see ¶204). In these situations the representative should first consult with the employer in an attempt to solve the problem.

The functions of a workplace health and safety committee set out at ¶508 also apply to a health and safety representative.

Training

Adequate training of health and safety representatives is crucial to the successful functioning of the role. Generally the legislation states that representatives must be granted time off without loss of pay to attend accredited training courses (see ¶204). Training courses are available through government departments, union bodies and private agencies (see the organisations listed in Chapter 12).

¶508 Workplace health and safety committees

Many organisations have found it beneficial to set up a health and safety committee at the workplace to investigate and discuss health and safety matters. Provision for such committees exists in the occupational health and safety legislation of all States and Territories (see ¶204). In New South Wales a health and safety committee is mandatory where the workplace has over 20 employees and the majority of employees request the formation of such a committee. In the other States health and safety representatives are given primacy by the legislation, but a committee may be established when requested by the representative or the employees.

Composition of committee

Ideally, the workplace health and safety committee consists of representatives of several levels of management, with at least 50% of the members being elected employee representatives. The inclusion of a senior management representative will give the committee status and authority as long as the manager has the authority to make decisions. A joint committee can operate effectively within any size of organisation. The size of the organisation, along with the number of separate departments within the organisation, may determine the size of the committee. Note that a committee of more than eight or ten people will tend to become unworkable. The legislation in most cases provides a guide to the composition of the committee, as well as the method of selecting the employee representatives (see ¶204).

Functions of committee

Information is basic to the effective functioning of a workplace health and safety committee. Committee members should have the right to inspect work premises and to obtain information concerning the safety of work products and processes (in some States this is enshrined in legislation (see ¶204)). The committee provides a mechanism for investigation and so matters coming under its scrutiny may be wide.

The functions, however, are advisory and educational, not decision-making. Management's ultimate authority, and that of supervisors, must not be undermined by the committee. Therefore, a committee can make recommendations to persons who do have the authority to implement them.

Typical aims of a workplace health and safety committee are:

- to enable rank and file workers to take an active part in the promotion of a healthy and safe workplace;
- to give employees the understanding necessary to assess the feasibility and cost factors of proposed risk control measures;
- to obtain the benefit of the great store of knowledge and experience possessed by many employees regarding health and safety aspects of the work they perform by obtaining feedback from them;
- to review measures taken to ensure the health and safety of persons at the workplace;
- to investigate matters brought to the employer's attention which a committee member or employee considers to be unsafe or a risk to health (resolution of that may involve recourse to an inspector, appointed under the relevant Act, to inspect the workplace);

¶508

- to assist in the development of recording systems for accidents and hazardous situations and to promote among employees an understanding of such matters as accident causation by reviewing recent accidents;
- to assist in the development of a safe working environment and safe systems of work;
- to assist in the formulation and effective implementation of an organisation's overall occupational health and safety policy;
- to monitor measures taken to ensure the proper use, maintenance and (if necessary) replacement of protective equipment; and
- to make recommendations to the employer regarding health and safety matters.

Advantages of a committee

A well-run workplace health and safety committee offers the following potential advantages:

- it helps to keep everyone informed on health and safety matters;
- interest in safety and health is widened by giving active participation to a number of people;
- the bringing together of varying viewpoints may result in a better decision than one developed from a single viewpoint;
- it can be an effective communication link between all department/section levels regarding a health and safety program, and this involvement promotes the sense of personal commitment and responsibility prerequisite to such a program's success;
- a preventive, rather than reactive, approach towards health and safety may be taken;
- management and employees are brought together — in the long run, better safety and health at work benefit both;
- it can provide an opportunity for frank, open discussion of problems; and
- it can obtain procedural information from employees who have knowledge and experience of on-the-job activities.

Against these advantages, there is the possibility that the committee will become simply a "talkfest", rather than a forum for actually generating improvements. There is also the problem (already mentioned at ¶507 in the context of health and safety representatives) that the committee will become entwined in general industrial relations issues. Both of these problems can

¶508

Staffing the Health and Safety Function

be overcome by having a strong chairperson and by setting agendas (see below).

Conducting a health and safety committee meeting

Timing of meetings should be arranged according to the organisation's own needs and workloads. They should be frequent enough to avoid an excessive number of topics having to be discussed at each meeting.

A suitable time for conducting meetings could be either one hour before lunch or one hour before finishing time, thus providing an automatic time limit on meetings.

Typical duties of committee members are set out below.

1. *Chairperson*: arrange for meeting venue and time, review previous minutes and materials for meeting, arrange program, notify all members of above items and make other arrangements as necessary.

2. *Secretary*: prepare and distribute minutes, prepare agenda, forward recommendations and deal with correspondence.

3. *Members*: attend all health and safety meetings, report all hazards and incidences of unreported injuries, accidents or near misses, contribute ideas and suggestions, work safely and endeavour to influence others, and carry out other duties as delegated.

Training

Here, as elsewhere, appropriate training must be considered (for example, for the development of the necessary skills of communication to participate in meetings and keep constituents informed; in preparing and conducting meetings; and to appreciate the aims and objectives of relevant legislation). Training courses are available from some of the organisations listed in Chapter 10.

¶509 The first aid officer

State legislation generally requires employers to provide first aid facilities in their workplaces. Industrial awards may contain a similar requirement. The scale of facilities required varies with the number of employees.

The relevant State legislation is:

NSW:

Occupational Health and Safety (First-Aid) Regulation 1989

Vic:

Code of Practice on First Aid in the Workplace — code of practice made pursuant to the *Occupational Health and Safety Act 1985*.

Qld:

Code of Practice for First Aid in the Workplace (made pursuant to the *Workplace Health and Safety Act 1995*).

SA:

Code of Practice for Occupational Health and FirstAid in the Workplace.

WA:

Occupational Safety and Health Regulations 1988, reg 327, 328, and the *Code of Practice for First Aid, Workplace Amenities and Personal Protective Equipment*.

Tas:

has released a draft *Code of Practice on First Aid in the Workplace*.

Regardless of legislation, employers should provide first aid facilities on a scale according to type of work and number of employees.

First aid facilities include:

- a properly stocked first aid box (as to contents, see below);
- a stretcher;
- a shower and/or eye-bath (where chemical hazards are present); and
- a first aid room (large organisations only).

Several employees should be trained in first aid, with one of them placed in charge of first aid facilities (including stock and maintenance). Many awards prescribe a first aid allowance to be paid for performing this duty.

The health and safety professional should regularly inspect the facilities and ensure that first aid officers hold an appropriate (and current) certificate. First aid courses are conducted by both the Australian Red Cross and the St John Ambulance Association.

The names and locations of the current first aid officers should be well publicised throughout the organisation.

Provision of first aid facilities should be regarded as a minimum requirement, even if the organisation has no other form of occupational health facilities or service.

Staffing the Health and Safety Function

In an organisation that has its own occupational health service, first aid facilities should still be available at appropriate locations in case of emergencies or very minor injuries that do not require the attendance of occupational health staff and to cover work times when OHS service staff are not in attendance, eg shiftwork, weekends.

Occupational health staff should supervise the first aid staff in charge of these facilities and oversee the maintenance and stocking of first aid boxes, etc. First aid staff should still be appointed from the shop, office, construction site or factory floor.

General duties of a first aid officer
1. Dispense and control items from first aid cabinet.
2. Ensure cabinet supplies are adequate.
3. Treat minor wounds and injuries, such as applying dressings, stopping bleeding and treating burns.
4. Deal with fits/fainting.
5. Resuscitation.
6. Record accident/injury details in accident book.
7. Arrange further assistance if required.
8. Advise top management immediately of any serious or potentially serious accident for which treatment has been required.

Contents of first aid box

Advice on first aid facilities may be obtained from State Health Departments. Legislation and regulations in each State should be consulted, as in some cases a list of minimum contents is set out. At the same time, there may be peculiarities of the individual work location which mean that extra items are needed. This information can be obtained from the health and safety survey.

OCCUPATIONAL HEALTH AND SAFETY PROFESSIONALS

¶510 When to engage an expert

Larger organisations will wish to employ health and safety staff as part of an on-site health service. The most common health staff on-site are occupational health nurses and occupational physicians. As on-site staff these practitioners will have the advantage of being familiar with the workplace, the types of safety problems encountered, which chemicals are used at the workplace, etc.

The duties that can be undertaken by on-site health nurses and doctors are set out at ¶511 and ¶512. Further commentary on occupational health services can be found at ¶714 et seq.

The need for and role of such employees will be affected by the type of workplace, type of work performed, particular safety aspects, availability of community health facilities, shiftwork and other factors. If occupational health practitioners are appropriate the following provides general guidelines on staffing recommendations:

1. Factories with over 300 employees, or shop/office with over 500 employees — one occupational health nurse and a part-time doctor.

2. Over 1,000 employees — at least two occupational health nurses and a part-time doctor.

3. Over 2,000 employees — at least one full-time doctor in addition to the nurses.

4. Over 4,000 employees — at least two doctors and at least three nurses.

Consultants

Special problems and staff limitations may give rise to the need to engage an outside expert. Using consultants has the advantage of bringing in more specialised skills than the organisation generally has access to. Consultants are separate from the workplace and are able to provide a fresh approach and an objective assessment.

Remember that when the outside assistance needed is more in the nature of general advice rather than a detailed assessment, various government authorities may be able to help. The major organisations are listed in Chapter 12.

Staffing the Health and Safety Function

Engaging a consultant

Preparation prior to engaging a consultant can save time and money. Remember that you are a consumer and the consultant is offering a service.

The following steps can be adopted:

1. Define the problem and what the consultant will be required to do.
2. List those consultants who may be able to assist.
3. Contact the consultants by phone and compile a short list of those who are suitable.
4. Call for written proposals from those on the short list.
5. Evaluate the proposals — cost is not the only criteria.
6. Draw up a contract specifying the exact nature of the project required — where the project is lengthy specify interim reporting requirements and the ability to alter the project.

¶511 The occupational health nurse

An occupational health nurse is a qualified nurse who has undertaken specific studies in occupational health. The Australian Occupational Health Nurses Association has issued a set of guidelines for the employment of occupational health nurses in Australia. The following outline is based on material in that publication.

Duties

The duties of each occupational health nurse will vary according to the size of the organisation, the number of employees, the nature of the work and associated hazards, the location and, of course, the number of nurses and other staff employed in the health service. These duties are all directed towards the prevention of occupational illness and injury and the promotion of physical and mental health.

See the list on the following page. It provides an indication of the range of duties.

The trend in the role of occupational health nursing in recent years has been to move more towards preventive functions. Many nurses will be involved in designing and implementing workplace programs in addition to the more traditional "band-aid" role.

Duties of an occupational health nurse

1. Pre-placement health assessment of new employees or employees changing their job within the organisation.
2. Monitoring employees' health routinely, for statutory purposes or for those at risk (eg executives, employees exposed to noise).
3. Assessment of employees who return to work after illness or injury, or who request permission to leave the premises because of illness.
4. Surveillance of employees who need special attention, such as those with known medical conditions, pregnant women, young people, persons with disabilities and workers exposed to special risks, such as lead.
5. Provision of immediate and follow-up nursing management in cases of illness or injury occurring at work.
6. Referring employees where appropriate to their own general medical practitioner, hospital, specialist or the company's medical adviser.
7. Giving advice to individual employees on health and hygiene.
8. Organisation of first aid facilities and services.
9. Supervision and training of first aiders.
10. Establishment and maintenance of accurate, complete and confidential health records, that may include information for workers compensation requirements.
11. Preparation of appropriate statistical records of all attendances at the health centre.
12. Preparation of reports for management including in-plant assessments and nursing activities.
13. Involvement in risk management programs, safety surveys, work environment visits, and safety and health education activities.
14. Liaison, cooperation and communication with other personnel, departments, trade unions, community organisations and programs concerned with employees' health, safety and welfare.
15. Health and safety research.
16. Involvement in counselling and rehabilitation.
17. Recommendation of other OHS specialists who may be appropriate.

Qualifications and training required

In addition to basic training and some practical experience as a registered nurse, the occupational health nurse should acquire a knowledge of occupational health nursing principles and practice, and keep up to date with developments in the field. As in other areas, new courses are becoming available that assist occupational health nurses to extend their skills.

¶511

Staffing the Health and Safety Function 101

Knowledge and skills in the following areas will be required:

- occupational diseases, epidemiology and toxicology;
- treatment of conditions such as eye injuries, burns, skin problems, strains and sprains;
- health screening procedures, including health interviews;
- counselling and communication skills (oral and written);
- training and instructional skills as needed to institute specific programs;
- selected laboratory procedures for biological and workplace environment monitoring (for example, to assess the need for referral to a specialist adviser);
- awareness of local community health and specialist occupational health and safety services available;
- local health and safety and workers compensation legislation;
- ethics (including impartiality) and privacy issues; and
- industrial hygiene.

¶512 The occupational physician or company doctor

The promotion of occupational medicine as a specialist branch is being recognised and encouraged by an expansion in the availability of appropriate training and in professional bodies (such as The Australasian Faculty of Occupational Medicine (AFOM) listed in Chapter 10).

It is still unusual for an occupational physician to be engaged outside of large organisations. More often an organisation will employ the services of a local general practitioner on a part-time basis.

Some general practitioners are now specialising in occupational health, and the organisation should "shop around" before choosing a part-time doctor (qualifications and training of occupational physicians are discussed below). Specific knowledge of the area in which the organisation operates is a distinct advantage.

The main duties of a "company doctor" can only be outlined very generally in the following table. Many of these services would involve further referral.

Duties of a "company doctor'

1. Liaison with and supervision of the work of the occupational health service staff, including nurses.
2. Advice on policy matters.
3. Treatment and consultations.
4. Medical examinations, including pre-placement.
5. Workers compensation cases.
6. Counselling and referrals.
7. Visits and investigation of work environments.
8. Research program development.
9. Involvement in health education programs.
10. Overall monitoring of the health service's performance, including reports to management.
11. Diagnoses.
12. Prescriptive and dispensary functions.
13. Surveillance of hygiene standards, including facilities such as canteens, kitchens, child-minding centres and sanitary installations.
14. Supervision of specialised medical equipment and its use.
15. Involvement in adaptation of jobs to employees from a safety/health viewpoint, including rehabilitation programs and disabled employees.
16. Liaison with managers, and also with outside local medical services.

The occupational physician or company doctor should report directly to senior management, although it is crucial for him/her to maintain a stance of strict impartiality (that is, to avoid becoming regarded as a tool of either management or employees).

A company doctor will be of greatest benefit to both the organisation and employees if the duties of the position are interpreted to include surveillance of both the workplace (processes, materials and conditions) and workers. This in turn suggests it is far better for consultations to be carried out at the workplace than at an outside surgery.

Consequently, where a doctor is employed as a consultant to the organisation, it will be necessary to provide adequate medical care and consultation facilities at the workplace. Administration of these facilities would generally be the responsibility of an occupational health nurse, perhaps assisted by clerical or other staff.

¶512

Staffing the Health and Safety Function

Qualifications and training required

Ideally, the doctor should have some postgraduate training in the occupational medicine/health field. Among the extra areas of knowledge and skill required are the following:

- an appreciation of management styles and problems, organisational behaviour, cost/benefit analysis, industrial relations and administration;
- familiarity with what employees do in their day-to-day work, including materials, plant and processes involved; and
- knowledge and experience in areas such as epidemiology, toxicology, ergonomics, safety, sanitation, health/safety legislation, health education, rehabilitation and counselling.

Under certain State legislation, provision is made for medical officers to become registered medical practitioners for specific health hazards, such as lead or asbestos.

Where particular hazards are involved, the doctor should have a thorough knowledge of the specific hazard and the appropriate treatments available, although detailed research study of the hazard may be undertaken by other service groups.

¶513 The occupational hygienist

Occupational hygiene may be defined as the detection, evaluation and control of either physical or toxic stresses in the working environment, that may have adverse effects on employee health and wellbeing. It aims to prevent adverse effects occurring, both in the short and long terms.

An occupational hygienist would be of particular value if the organisation has a health problem that needs constant monitoring, such as the use of a potentially harmful chemical or substance. Many of these are of unknown potential, so it is essential to keep levels of exposure to the lowest practical limit.

Duties of an occupational hygienist could include those in the following list.

As this list implies, the hygienist will need to possess knowledge of hygiene, work study methods, sanitation, general health, research techniques and communication skills. Frequent contact with the workplace and its employees is essential, as is the need to keep abreast of new discoveries and developments, particularly the long-term effects of hazards and control measures.

The hygienist could be either a full-time employee or an outside consultant.

Duties of an occupational hygienist

1. Establish standards of exposure or threshold limit values based on known effects of particular hazards, and use them as guides to control measures.
2. Audit the work processes/materials that involve these hazards and develop monitoring programs to evaluate the extent of the problem (note that too fast a pace of work can be a hazard).
3. As a result of 2, advise management on control measures (by means of risk management processes).
4. Work with industrial medical staff to evaluate particular tasks, processes or materials.
5. Prescribe measures for safe handling and treatment of chemicals.
6. Enable management to provide advice on hazards to employees, safety personnel, compensation authorities, customers and governments.
7. Provide data for continued surveillance (that is, whether the situation is improving or worsening).

¶514 Other occupational health and safety consultants

It may be that an organisation requires the assistance of a consultant with expertise in a specific area. The occupational health and safety professionals presented in this paragraph cover the major groups not already discussed. Because of the specialist nature of their work, all but the largest companies would be expected to engage these professionals on a consultancy basis.

Organisational psychologist

Organisational psychologists are concerned with people at work. They apply psychological principles to the study of work performance, work attitudes, working conditions and organisational structures. They aim to improve individual performance and work satisfaction, as well as promoting organisational effectiveness.

Areas of occupational health and safety with particular relevance to a psychologist include employee assistance programs, occupational rehabilitation, stress and general health maintenance.

The Australian Psychological Society is the professional body monitoring standards and representing psychologists. There is a branch and also a Board of Organisational Psychologists in each State.

Risk manager

Risk management addresses not only employee health and safety, but all elements of unnecessary costs that may affect a company's profitability, eg loss prevention, computer security.

Risk management involves minimising risk through appropriate management systems. Risk managers may look at the whole organisation, however with respect to OHS this involves identifying hazards/risks and developing systems to minimise the risks associated with those hazards. It is a discipline oriented to practical change. See also Chapter 3.

The risk manager:

- identifies the hazards associated with a particular undertaking;
- qualifies and quantifies the degree of risk associated with identified hazards (what potential these have to cause human and financial injury/loss);
- designs viable programs to limit exposure to the hazards/risks (so reducing the potential for injury/loss); and
- monitors and maintains these programs.

Rehabilitation counsellor

A rehabilitation counsellor provides assistance to people with problems resulting from disability or injury. In the occupational context this would refer to people who are having problems getting into, or returning to, the workforce. The counsellor may assist in the process of adjusting to the workforce through individual counselling or through coordinating various other services.

A rehabilitation counsellor will be concerned with the whole person, not simply with the person's disability. He/she may stress the remaining skills of the injured person, or may assist the worker in obtaining new skills.

Rehabilitation counsellors may also be involved in assessing the effectiveness of any return to the workforce.

Ergonomist

Ergonomics looks at people and the systems in which they work. It is about designing work (including machines and workplaces) to suit people. The employees are considered to be the chief resource at the workplace, around whom work systems should be structured.

An ergonomist considers areas such as the following:

- equipment design (handles, position of controls, etc);

¶514

- workplace layout;
- environmental factors (light, noise, chemicals);
- how people handle information;
- the skills people use; and
- how individual variations between people (such as height, weight, strength, intelligence) can be accommodated by a work system.

Ergonomics is a very broad science and it is hard for any one person to claim to be an "ergonomics expert". Some consultants come to ergonomics via other professions such as engineering, physiotherapy or nursing. The Ergonomics Society of Australia provides accreditation for its members. The National Occupational Health and Safety Commission has a large amount of information on ergonomics on its website (www.nohsc.gov.au).

¶515. Establishing a group health service

Small organisations could give thought to either maintaining a part-time service or combining with other small organisations within close proximity to establish a group health service. An obvious example is an industrial estate complex. If establishing a group health service, either a central health centre, a series of separate clinics, or a mobile centre could be considered. Several such casual, on-site services are now available on a commercial basis, offering a range of services from pre-placement testing and specific services (such as audiometric testing) to ongoing monitoring.

The suitability of any such approach will depend on the availability of nearby medical/health facilities and the needs of each organisation. It may also require frequent reviews to make sure it is continuing to meet those needs.

Problems that may have to be addressed with a group service may include greater complexity of administration (with several companies participating), a wider variety of existing and potential problems and hazards, and financial support (each company would normally pay a "subscription" to the service). Financial planning in advance will be very important (for example, to establish an optimum size, range of services and coverage of companies) to ensure financial viability.

Chapter 6

Techniques of Accident Investigation, Prevention and Reporting

Introduction .... ¶601
What is an accident? .... ¶602
Accident causes .... ¶603
"Accident proneness" and "blame" .... ¶604
Common types of accidents .... ¶605
Accident prevention flowchart .... ¶606
The steps of accident investigation .... ¶607
What should be investigated? .... ¶608
Who should investigate? .... ¶609
Investigating accident causes .... ¶610
Special equipment .... ¶611
Accident recording systems .... ¶612
Accident reporting requirements .... ¶613
Statistics .... ¶614
Report writing .... ¶615
Preventing recurrence of accidents .... ¶616
Feedback to employees .... ¶617
Disciplinary policy .... ¶618

¶601 Introduction

The purpose of this chapter is to examine some of the techniques commonly used in accident investigation and some of the means of monitoring progress.

The costs and effects of accidents and exposure to hazards at the workplace have already been emphasised in Chapter 1. In Chapters 7, 9, 10 and 11 implementation and the various components of a health and safety program are considered.

An essential part of a risk management program is to investigate accidents and near-miss incidents, so that the causes can be addressed and the chances of the same or similar incidents recurring can be eliminated or at least reduced. This is another way in which hazards can be identified and risks assessed and controlled, and it requires both a systematic means of investigation and the maintenance of a record system to monitor progress.

¶602   What is an accident?

As defined at ¶109, a workplace accident may be described as any event arising out of employment that results in human injury or damage to property, or the possibility of such injury or damage.

For the purposes of accident prevention, however, it is also important to record and investigate all "near-miss" situations (that is, events which did not result in injury or damage but that had the potential to do so).

An accident or near-miss will be preceded by an unsafe act and/or an unsafe condition. This is discussed under "Accident causes" at ¶603 below.

¶603   Accident causes

The complex nature of accidents has already been referred to in Chapter 3. All accidents or near-miss situations represent a failure in, or an inadequate, OHS management system at the workplace. All accidents or near-miss situations are preceded by:

(a) an *unsafe act* — an act by the injured person or another person (or both) that caused the accident; and/or

(b) an *unsafe condition* — some environmental or hazardous situation that caused the accident independent of the employees(s).

Identifying the real cause of an accident is, however, a complicated process. Investigation of an accident may reveal several coincidental causes that create a chain of causation factors, none of which would have resulted in an accident by themselves.

The basic concept of accident prevention is that accidents can have several causes, each of which can be identified and controlled.

For example, an unsafe act may have been due to the lack of a defined company policy, or the lack of proper training, or faulty supervision, leading to a breach of rules. Alternatively, it may have been due to risk taking or skylarking. Another cause of accidents could be costcutting, such as lack of adequate maintenance or use of cheap, unsatisfactory protective equipment.

The following examples indicate the complexity at times of the process of identifying causes.

Techniques of Accident Investigation, Prevention and Reporting

1. An employee slips on an oil patch in a warehouse.

Is it because of:

(a) bad housekeeping (that is, no one wiped up the spilt oil, which could be due to inadequate information, training or supervision; or the absence of documented safe working procedures);

(b) poor maintenance (if the oil was dropped by a fork-lift truck); and/or

(c) faulty management (if the transport officer is on leave and no one renewed the maintenance contract)?

2. A truck driver ignores signs of brake failure and is injured

When a truck driver ignores a sign of brake failure and is injured when the truck fails to stop at an intersection and collides with a car, the accident is due to the driver's failure to respond to the danger signal. One should therefore look for reasons why the driver chose to ignore it, instead of dwelling solely upon the accident itself. Was it because:

(a) the delivery schedule was too tight;

(b) the driver was over-confident or inadequately-trained (eg driver thought the sign of brake failure was within tolerable limits); and/or

(c) it was the last load for the day?

The implication of these examples is the need to focus on all the links in the chain of events that led to the unsafe acts or conditions, rather than the consequences (that is, the injury or damage).

Among the most common underlying causes are the following general categories:

- lack of attention to health and safety procedures and practices (such as failure to assess manual handling risks or failure to replace a guard);
- employees' failure to use health and safety equipment when supplied, such as personal protective clothing (see ¶629);
- use of poor quality health and safety equipment;
- inadequate training, instruction and supervision of employees (or failure to understand instruction), even at the original induction stage of employment;
- management's neglect of its common law obligations to provide a safe system of work, a safe place of work, competent staff, a sufficient number of workers and proper plant and equipment;
- inadequate attention to maintenance and "housekeeping"; and
- failure to provide adequate safeguards and/or suitable equipment for potentially dangerous situations.

¶603

Attention should also be paid to psychological factors. For example, if an employee finds a job boring, monotonous, or well below his/her capabilities, the result may be alienation from the job, emotional stress, lack of care and loss of concentration, all of which increase the chances of accidents occurring. The solution to these problems will involve an examination of job design and the match of employees' capabilities to suitable jobs.

Job demands must also be considered, such as the effects of long working hours, shiftwork, deadlines, piecework schemes which emphasise increased production at all cost, and other physical and mental stresses.

¶604 "Accident proneness" and "blame"

Current approaches to occupational health and safety avoid making value judgments as to whose "fault" caused an accident or who can be "blamed". This is reflected by no-fault workers compensation systems. Instead, the system aims at investigating and explaining accidents with a view to identifying the fault in the work system that led to the accident, and identifying control measures by which accidents can be avoided.

In the same way, the concept of identifying certain people as being "accident prone" has been rejected by social scientists due to the lack of evidence of any specific personality characteristic that makes a person susceptible to accidents. Focusing on only one cause or person tends to obscure the whole chain of events that may contribute to the accident.

It is possible, however, for employees to become more prone to accidents in time because of undetected, normal deterioration of eyesight, hearing, body movements, and/or the use of medication. It is also essential to be alert for "repeaters" in relation to type of injury or time of year, such as recurring back problems, heatstroke or unsuitable training of seasonal workers.

¶605 Common types of accidents

Workers compensation statistics suggest that the following are the most common types of accidents:

- manual handling injuries/over-exertion or over-strenuous movements;
- stepping on, striking against or being struck by objects; and
- slips, trips, falls and jumps by persons.

Each of these causes can be addressed by risk management, effective health and safety training practices and good supervision.

¶606 Accident prevention flowchart

As mentioned above, an effective risk management program should aim to stop accidents from occurring in the first place. The flowchart on the following page covers both steps towards preventing accidents and a follow-up procedure if an accident still occurs, and shows the interrelation between components of an accident prevention program.

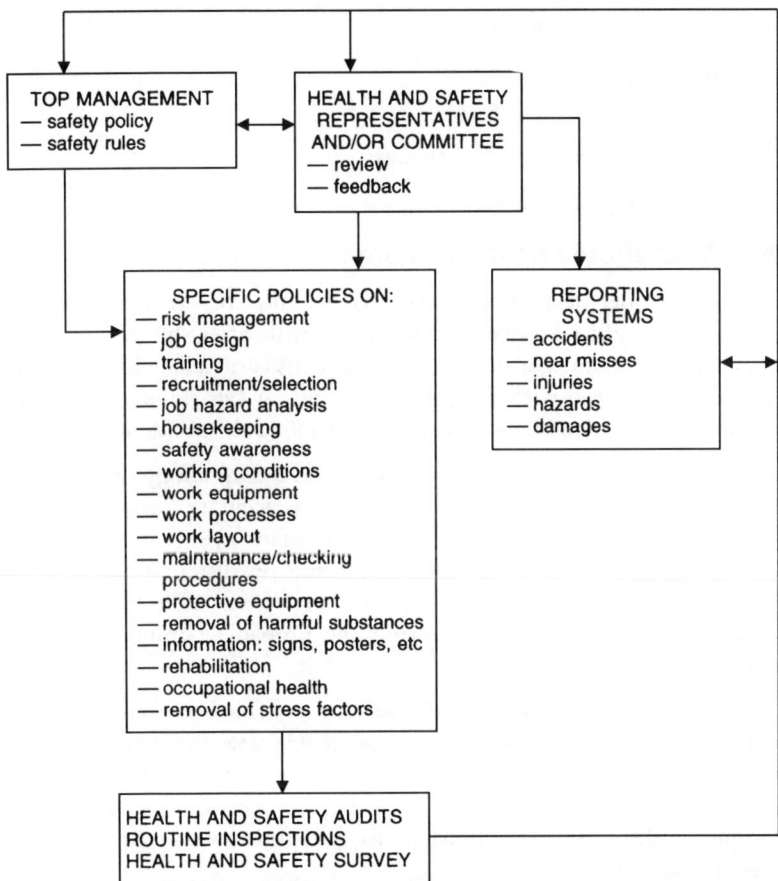

¶607 The steps of accident investigation

As with other management activities, accident investigation requires certain defined steps to be taken by nominated people in terms of procedures, use of resources, paperwork and follow-up action. These steps, listed below, are discussed in the remainder of this chapter:

- what should be investigated (¶608);
- who should investigate (¶609);
- special equipment (¶610);
- investigating accident causes (¶611);
- accident recording systems (¶612);
- accident reporting requirements (¶613);
- accident statistics (¶614);
- report writing (¶615);
- preventing recurrence (¶616);
- feedback to employees (¶617); and
- disciplinary policy (¶618).

¶608 What should be investigated?

All accidents or near-miss situations should be investigated. This investigation should take place as soon as possible after the incident occurs. Promptness of investigation is important as crucial evidence can be disturbed or destroyed with the passage of time. A camera is a useful tool as it captures conditions as they existed at the time of the accident.

Both a survey of the accident location and a chronology of the accident events should be obtained. This will involve accurate measurement and probably the preparation of a sketch map. As many eyewitnesses as possible should be questioned. The observations made by different witnesses will probably conflict so it is better to concentrate on collecting witnesses' reports on the spot and reconstructing the probable chronology after the initial questioning.

Accident investigation should not be confined to on-site analysis. Statistics and past trends may indicate particular areas for concern involving certain work processes, injury types, time losses, groups of employees, or locations. Details of the investigation should be conveyed to similar work situations within the organisation to promote awareness of possible undetected hazards there.

¶609 Who should investigate?

Each accident or near-miss should be investigated by the immediate supervisor of the section concerned, although assistance may be sought from others. Selected incidents may be separately investigated by the health and safety officer/manager, occupational health nurse, and health and safety representatives and/or committee. Note that anyone who carries out an investigation should have the training to do so.

Techniques of Accident Investigation, Prevention and Reporting

The following is a list of desirable skills that the persons in charge of the investigation process should ideally possess:

- a range of basic investigative skills, recording skills (photography, drawing of plans) and technical knowledge (of work processes, equipment and work layout);
- ability to interview people, such as the injured person, witnesses and others with relevant knowledge;
- a capacity to understand the needs and reactions of employees;
- skill in report writing;
- ability to observe and note details (for example, when attending the accident site); and
- ability to recommend preventive action, and implement it where it is within the person's authority to do so, and assess the effectiveness of such action.

It is advisable that more than one person be present at the on-site accident investigation. Often an outside authoritative source may need to be called in, immediately if possible, or at a later date when the work task or process is in progress again.

¶610 Investigating accident causes

Carrying out an investigation aimed at identifying the underlying causes of an accident or near-miss will require asking some basic questions. The importance of concentrating upon the underlying cause, rather than the outcome, was emphasised at ¶603. The implications are that it will usually not be sufficient to interview the employee(s) directly involved. In addition, an examination of the workplace and its operation methods will be required, as well as the collection of background information.

If a person is injured carrying out a particular task, it is necessary to establish whether he or she considered alternative ways of doing that task. This is important because if the person knew no safer way of doing the task, then the most appropriate form of remedial action may be training or instruction. If, however, the person did know of a safer way and chose, for whatever reason, the more dangerous way, then training is unlikely to be an effective type of remedial action and other means of addressing the causes will need to be considered.

One useful question to ask employees during an interview is something along the lines of: "What do you suggest to prevent a future recurrence of this accident/incident?". This may be an appropriate subject for review by a workplace committee.

> The primary objective of accident investigation from an organisational and OHS authority perspective is to prevent future accidents, not to assign blame. This should be stressed to employees who are interviewed following an accident, to ensure that they provide all relevant information. Government authorities, however, may need to investigate incidents for purposes of initiating prosecution.

¶611 Special equipment

Equipment required for investigating accidents or incidents will of course vary according to the type of workplace, but the following table suggests a basic range of equipment that could be useful. A kit should contain measuring, recording and marking devices, with each item readily accessible and ready for immediate use.

> **Camera and film** (preferably VHS with play-back, digital or polaroid to allow for immediate assessment of the adequacy of the photo).
> **Cassette tape recorder** for witness interviews, oral notes, observations and reports.
> **Sound level meter** to measure and analyse noise levels.
> **Gas and vapour analyser** to measure carbon monoxide and other gas vapour concentrations.
> **Electrical receptacle tension tester** to test electrical continuity and grounding of receptacles and distribution cords.
> **Wire/hemp rope calculators** to measure rope size for analysis of load capability of ropes used for lifting and stabilising.
> **Callipers** for precision measurement of parts: inside and outside dimension callipers as pertinent.
> **Sample containers** of type, material and size appropriate for collection of fuels, lubricants, coolants and other substances.
> **Protective equipment**: a range for use by investigators.

¶612 Accident recording systems

All employers should establish a system for recording workplace injuries, near-misses and diseases. The availability of reliable and usable data is essential for the development, monitoring and evaluation of health and safety strategies.

Information from an accident recording system can be used to complete statutory accident notification requirements (see ¶613), but its prime function is to initiate preventive action to control future incidents. With this in mind, any recording system must allow the identification of trends in the types of injuries, time of occurrences, location of occurrences and agents

Techniques of Accident Investigation, Prevention and Reporting

of injuries or diseases. The amount of time lost as a result of each injury/disease should also be recorded. Recording this information allows safety statistics to be compiled (see ¶614).

The National Health and Safety Commission, in conjunction with Standards Australia, has developed the *Workplace Injury and Disease Recording Standard (AS 1885.1:1990)*. The Standard provides employers with a simple method of recording comparable data covering accidents, near-misses and diseases at the workplace. It can be used by both large and small employers. The sample recording form, which is published as an appendix to the Standard, is reproduced on the following pages. Note that it is not necessary to use the sample form in order to implement the Standard. Some organisations may wish to design a form with additional information.

Medium and large organisations may want to computerise their workplace accident and disease records. Standards Australia offers a software package that allows users to record all necessary information on workplace injuries in a form that fully complies with *AS 1885.1*.

Use of *AS 1885.1* will satisfy requirements contained in State health and safety legislation regarding the recording of injuries. An example of a Workplace Injury and Disease Recording Form is on pages 116 and 117.

AS 1885.1—1990

WORKPLACE INJURY AND DISEASE RECORDING FORM

Reference number [ ]

Personal details of the injured worker

1. Surname [ ]
   Given names [ ]

2. Sex (M or F) [ ]

3. Date of birth Day / Month / Year

Basis of employment

4. Starting time:
   1. [ ] 0600 – 1159 hours
   2. [ ] 1200 – 1559 hours
   3. [ ] 1600 – 0559 hours

5. Shift arrangement:
   1. [ ] Fixed, standard or flexible hours
   2. [ ] Rotating shift

6. Number of hours:
   1. [ ] 8 hours or less
   2. [ ] more than 8 hours (excluding overtime)

Job details

7. Description of occupation or job title [ ]

8. Main tasks performed [ ]

9. Training provided:
   1. [ ] Induction training
   2. [ ] Task specific training
   3. [ ] Both of the above
   4. [ ] Neither of the above

Details of the injury or disease

10. Date injury occurred or disease reported Day / Month / Year

11. Time injury occurred [ ] [ ] (24 hour clock format)

12. Nature of injury or disease [ ] Code [ ]

13. Bodily location of injury or disease [ ] Code [ ]

14. Description of occurrence of injury or disease:

   • In which part of the workplace did the injury or disease exposure occur?
   (e.g. machine shop, freezer room, No. 2 mine)
   [ ]

   • What was the worker doing at the time?
   (e.g. driving a fork lift truck, lifting bags of cement, typing)
   [ ]

   • What happened unexpectedly?
   Include the name of any particular chemical, product, process or equipment involved.
   (e.g. brakes failed on fork lift truck, slipped on wet floor, scaffolding collapsed, arm started hurting while typing on a word processor)
   [ ]

   • How exactly was the injury or disease sustained?
   Include the name of any chemical, product, process or equipment involved.
   (e.g. hit head on cabin of fork lift truck, lacerated knee when landing on ground, arm hurt after long period of typing)
   [ ]

   (See pages 25 – 28 of the Standard)

   Code
   Mechanism of injury [ ]
   Breakdown agency [ ]
   Agency of injury [ ]

¶612

Techniques of Accident Investigation, Prevention and Reporting

AS 1885.1—1990

Lost-time injury/disease

Additional questions to be answered for cases which result in a fatality or permanent disability, or where there was time lost from work of one or more days/shifts. These questions should be completed as soon as possible after the injury or disease is reported.

15. Employee's preferred language [_____]

16. Type of employment:
   - 1 ☐ Full-time permanent
   - 2 ☐ Full-time casual
   - 3 ☐ Part-time permanent
   - 4 ☐ Part-time casual

17. Type of employee:
   - Wage/salary earner:
     - 11 ☐ Trainee
     - 12 ☐ Outworker
     - 13 ☐ Apprentice
     - 14 ☐ Pieceworker (other than Outworkers)
     - 15 ☐ Other

   (Note: most employees will fall into this category)

   - Self-employed: 20 ☐ (including contractors and sub-contractors)
   - Unpaid worker:
     - 31 ☐ Work experience
     - 32 ☐ Other

18. Worker's experience in task being carried out when injury or disease occurred: [Years] [Months]

19. Proportion of shift worked:
   - 1 ☐ 25% or less
   - 2 ☐ 26% – 50%
   - 3 ☐ 51% – 75%
   - 4 ☐ 76% – 100%
   - 5 ☐ Overtime

Details of person completing this form

Name	
Position	
Signature	
Date	/ /

Outcome of injury/disease

Questions 20 – 24 are about information that is not available at the time of the report of injury or disease. These questions should be answered as soon as the information becomes available. For some occurrences, such as where there was no time lost, some of these questions will not be relevant.

20. Rehabilitation:
   - 1 ☐ Required
     - ↳ Date of commencement of rehabilitation program
       - Day Month Year [/ /]
   - 2 ☐ Not required

21. Was the injury or disease:
   - 1 ☐ Fatal
   - 2 ☐ Non-fatal

22. Preventive action proposed or taken:

[_____]

(Tick one or more boxes as appropriate)

	Proposed	Taken
Change to induction training	11 ☐	12 ☐
Change to ongoing training	21 ☐	22 ☐
Equipment/machinery modifications	31 ☐	32 ☐
Change to work procedures	41 ☐	42 ☐
Change to work environment	51 ☐	52 ☐
Equipment/machinery maintenance	61 ☐	62 ☐
Other job redesign	71 ☐	72 ☐
Other preventive action	81 ☐	82 ☐

23. Date of resumption of work on:

	Day Month Year
Short-term alternative duties	/ /
Permanent alternative duties	/ /
Normal duties	/ /

(Enter each date when applicable)

24. Total number of working days lost [____]

(Should be completed only when the worker has resumed permanent duties)

¶613 Accident reporting requirements

State workers compensation legislation generally requires employers to keep a register of injuries, and in some cases to report injuries to the relevant authority. The *Workplace Injury and Disease Recording Standard* referred to in ¶612 may not satisfy this requirement, as often the workers compensation legislation prescribes a specific form of register. The organisation will thus be required to keep dual records, with the register of injuries being used to support workers compensation claims.

In addition to the above, occupational health and safety legislation requires that certain types of accidents, usually those involving death or absence from work beyond a certain number of days, must be reported to the body that administers health and safety in that State.

¶614 Statistics

Australian Standard *AS 1885.1* (see ¶612) sets out the following statistical formulae for use in industry.

Incidence Rate: the number of occurrences of injury/disease for each one hundred workers employed.

$$\frac{\text{number of occurrences in the period}}{\text{number of workers}} \times 100$$

The "number of occurrences in the period" refers to all injuries/diseases that resulted in lost-time of one day/shift or more during the specified period.

The "number of workers" is defined as the average number of workers who worked in the recording period. Persons who were absent from work on paid or unpaid leave for the entire period should be excluded from this calculation.

Frequency Rate: the number of occurrences of injury or disease for each one million hours worked.

$$\frac{\text{number of occurrences in the period}}{\text{number of hours worked in the period}} \times 1{,}000{,}000$$

The "number of occurrences in the period" is defined in the same way as for the incidence rate.

The "number of hours worked in the period" refers to the total number of hours worked by all workers in the recording unit including, for example, overtime and extra shifts.

Average Time Lost Rate: the average time lost per occurrence of injury/disease. This rate provides a measure of the severity of the occurrences being experienced by workplaces over time.

$$\frac{\text{number of working days lost}}{\text{number of occurrences in the period}}$$

The "number of occurrences in the period" is defined in the same way as for the incidence rate.

The "number of working days lost" refers to the total number of working days, irrespective of the number of hours that would normally have been worked each day, that were lost as a result of the injury/disease up to a maximum of 12 months for any individual occurrence. For the purposes of calculating the average time lost rate, occurrences that result in a fatality should be assigned a time lost of 12 months (220 standard working days).

It is recommended that medium and large organisations prepare these statistics every month, while smaller organisations may only need to calculate rates on a six-monthly or annual basis.

The main function of these statistics is to measure trends over time so that the organisation can tell whether its record is improving, stable or deteriorating. For this reason it is important that data be retained for several years. The statistics also provide a guide to the effectiveness of any corrective action taken as part of a health and safety program.

Converting information to chart form, particularly monthly figures over a 12-month period, can be useful in highlighting periods in which more accidents/injuries occur than in others. It may also be worthwhile to use a chart to compare accident/injury rates with absenteeism statistics, to check for similarities between the two.

In addition to these basic statistics, there are other calculations that might prove very useful in identifying particular "trouble spots" within an organisation. For example, accident statistics could be isolated and classified by department, type of work process, agent of injury/disease, age group of employees, length of service, length of time off work, type of injury and other relevant classifications. One advantage of such an approach is that "good" and "bad" sections of an organisation can be compared and, if relevant factors can be found, it may sometimes be possible to transfer the features of the "good" sections to other sections.

¶614

Caution on the use of statistics

It is important, however, that statistics be used with caution. Comparisons with other organisations will often be invalid, due to different work processes, employee backgrounds and so on. They could be difficult to obtain, as many firms would be reluctant to disclose such information.

For the purpose of communicating with employees, a simpler form of statistical presentation may be more effective. For example, many organisations use a notice headed "number of production hours without a work accident" or else set obtainable "targets" along similar lines, such as a five percent injury reduction each year.

It is essential to ensure that if the above types of notices or targets are used, they must not result in suppression of accident reporting. It has often been noted that such notices can create a social pressure that leads to under-reporting, and is therefore counter productive.

¶615 Report writing

The skill of report writing tends to be an underrated aspect of most managerial and supervisory functions. It is desirable to provide health and safety staff and all managers and supervisors with some training in this field. It will assist them to provide relevant information and express themselves clearly.

Firstly, to convince management that there is a problem and to spend money on the remedial action proposed, it will be necessary to present facts in a clear and logical fashion and demonstrate how the proposed solution will be the most effective or most cost-effective. Ideally, alternative solutions should also be suggested where feasible. Secondly, a report can provide a check on the adequacy of accident investigations, raising questions as to whether additional inquiries should have been made, or whether there is enough information to form definite conclusions. Further investigation may be necessary after this feedback before remedial action can be decided on.

Finally, the degree of loss or potential loss incurred by an accident should be quantified in monetary terms. A simple costing of the fundamental areas affected will provide management with an idea of the cost involved. The direct and indirect costs of an accident are noted in the diagram at ¶103.

¶616 Preventing recurrence of accidents

It must be remembered that the main purpose of investigating accidents is to prevent their recurrence. It is possible that an accident or incident will reveal the existence of hazards or shortcomings in the management system that were not previously identified by inspections, the safety survey or other efforts to ensure all risks were identified. If this is the case, these new insights should feed directly into the risk management system described in Chapter 3.

¶617 Feedback to employees

As mentioned earlier, a successful health and safety program will require employee awareness of issues, as well as their support and motivation both to work safely and to assist with the reporting and investigation of unsafe situations. This commitment will only be achieved if employees and their representatives can be convinced that management is acting in the employees' interests. This in turn will require constant feedback to employees on accident investigations and any changes that result from an investigation.

Information that could be supplied to employees includes the following:

- the physical results and costs of accidents to both employer and employees;
- the causes of an accident, giving both employer and employee versions if these differ;
- progress reports on any long-term projects and changes being undertaken; and
- statistical information on a regular basis, perhaps in chart or graph form for ease of understanding.

Management will need to follow up on the progress of corrective action as well. The timing and responsibility for doing this could be established from the accident report or from other records. Follow-up actions on the health and safety audit or surveys should be conducted periodically and not treated as "one-off" exercises.

¶618 Disciplinary policy

There are times when health and safety matters will require the disciplining of employees. This is a serious matter that should not be undertaken merely so that someone can be seen to be taking the blame for an accident.

Employees have legal responsibilities under OHS legislation and, if an employer condones an employee's breaches under that legislation, the employer is in effect in breach of the legislation as well.

A disciplinary policy should aim to alter the employee's behaviour in a positive way, rather than to "punish" the employee. The policy needs to be consistent and have the cooperation of employees. Factors to consider in the policy include actual and potential money costs, actual and potential injuries, related circumstances (such as equipment and working conditions), concealment by anyone of the occurrence, the employee's past record and treatment of any similar previous cases.

Other relevant factors include the employee's contract of employment and the organisation's rules on other matters.

Chapter 7

Implementing a Program

Introduction	¶701
Health and safety policy, program and plans	¶702
Sample health and safety policy statement	¶703
Health and safety procedures, rules and work method statements	¶704
Recruitment and selection of employees	¶705
Health and safety training	
Why train?	¶706
Induction training	¶707
Continuing training	¶708
Communicating health and safety	
Creating a high profile	¶709
Posters	¶710
Printed matter	¶711
Videos	¶712
Incentive schemes and competitions	¶713
Occupational health service	¶714
Rehabilitation	
What is occupational rehabilitation?	¶715
A rehabilitation program	¶716
Features of occupational rehabilitation	¶717
Rehabilitation systems — State by State	¶718
Safety and health off the job	¶719
Medical examinations	¶720

¶701 Introduction

This chapter assumes that the hazards associated with the work carried on by the employer have been identified, the risks arising from these hazards assessed and controls put in place (see Chapters 3, 4 and 6). The context of these activities is the infrastructure to enable the organisation to create and maintain a safe and healthy work environment.

The action required to achieve an appropriate infrastructure will include attention to a wide range of issues, such as the following:

- issuing a health and safety policy and endorsing health and safety programs and plans as required;
- on-going review and monitoring of the effectiveness of risk control measures and other elements of the health and safety program;
- health and safety instruction and training;
- establishment of performance indicators;
- promotion and communication of safety and health;
- establishment and administration of an occupational health service;
- establishment of a rehabilitation program; and
- medical examinations.

All of these aspects are discussed in detail in this chapter. The establishment of health and safety staff, representatives and committees is discussed in Chapter 5. Accident investigation, prevention and record keeping are considered in Chapter 6. Certain common health and safety problem areas are introduced in Chapters 9, 10 and 11.

The health and safety program will initially be based on the risk management process described in Chapter 3, but will alter as each level of improvement is reached. Achievement of the program's objectives depends on the commitment of management and the workforce. This depends in turn on the parties being informed about the programs being considered and their active involvement in planning.

The lead-up involved in implementing a health and safety program is similar to that of producing a service program. Based upon a needs analysis, an achievable objective is set (to be reached within a given time) and a forecast made (which must contain costing, human resources and equipment usage). It does not have to be elaborate, simply well planned.

Attention should be paid to the attitudes, backgrounds (such as ethnicity, language, religious customs and education) and needs of employees.

Employees and any union(s) represented at the workplace should be consulted throughout the planning and implementation of any health and safety program as described at ¶204.

¶702 Health and safety policy, program and plans

Issue of a policy statement on occupational health and safety by top management is the logical first step of a safety/health program. The health and safety policy should be distinguished from specific programs or plans to implement the policy.

The policy should state that the organisation accepts responsibility for the safety and health of its employees and express management's goals, responsibility, accountability and participation in the safety/health function. The statement should aim to enlist the support of all staff and employees, and should express support for safety representatives and/or a safety committee. It should not create a situation where what is said cannot (or fails to) be implemented at all organisational levels, otherwise double standards may result.

A "team" approach between management and employees in the formulation of the policy statement will provide an opportunity for workers to relate to the statement and also help develop a relationship of mutual trust.

Content of statement of policy

1. Expression of management's objectives and intentions.

2. A set of general *guidelines* to the health and safety function (not to be confused with more specific procedures).

3. Acceptance of primary responsibility for the health and safety function by top management.

4. Expression of support for health and safety representatives and/or the workplace safety committee.

5. Outline of authorities and responsibilities of all employees at all levels.

The policy must be publicised, either by distribution to employees, display on notice boards or announcement. Where there are employees who do not speak or understand English well, it is important to provide multilingual translations or convey the information by other means, eg verbal explanation.

Policies versus programs and procedures

A distinction needs to be drawn between the role of a policy (as a general statement of intention and objectives) and a program (a set of detailed steps that how the policy is to be implemented) and procedures, which may include safety rules or work method statements for particular tasks. The health and safety program will describe the organisation's plans for addressing particular issues. For example, a program on hearing conservation may cover testing procedures, noise reduction and use of ear-muffs, and it may be supported by a set of specific instructions.

¶703 Sample health and safety policy statement

See the sample health and safety policy on page 127. This is an example of an organisation's policy statement, to show what items could be included. Alterations can be made to cater for each organisation's different circumstances.

In large organisations, it may be a good idea for line managers to countersign the policy statement, as it emphasises local responsibility and improves the status of these managers where safety and health matters are concerned.

¶704 Health and safety procedures, rules and work method statements

Risk control strategies often include the documentation of particular procedures to be followed. This may take the form of basic health and safety rules to guide behaviour, or work method statements relating to particular jobs/tasks.

All employees will need to be informed of the rules by both receiving them and by prominent displays in the workplace.

The rules should be basic, supplemented by more detailed operating instructions when the latter are needed.

Where possible, observance of the rules should be contained in an induction document as a condition of employment, with failure to observe them rendering an employee liable for further action, such as retraining or, where appropriate, discipline.

> **Health and safety policy**
>
> It is the company's policy that each of its employees shall be provided with a safe and healthy place in which to work. To achieve this policy, management will make every reasonable effort in the areas of hazard identification, risk assessment and control, as well as health preservation and promotion. These aspects of working conditions will be given top priority in company plans, procedures, programs and job instructions.
>
> In conjunction with this policy, a series of programs, procedures, and rules on specific individual health and safety matters will be prepared and issued.
>
> Health and safety at work is both an individual and shared responsibility of *all* employees. The following areas of responsibility are essential to the success of the policy:
>
> 1. *Top management.* Top management is required to actively pursue the goals set out in the first paragraph of this policy through the following approaches:
>
> (a) devising and administering a comprehensive safety and health program;
>
> (b) holding regular senior staff and supervisors' meetings to discuss health and safety performances; and
>
> (c) taking effective action to provide and maintain safe and healthy working conditions for all employees.
>
> 2. *Supervisors.* Supervisors will be held accountable for working conditions under their control and for the extent of information, instruction, training and supervision given on safety and health matters to employees. They are to provide the initiative and follow-up action to maintain this policy within their own sections.
>
> 3. *All employees.* Employees share responsibility for their own safety and that of their co-workers. The success of a safety and health program ultimately rests on the willingness of everyone to co-operate and work collectively with a "team spirit".
>
> The workplace health and safety committee/health and safety representative shall be supported by management and employees.
>
> Reducing work-related injury and disease, as well as related insurance costs, will permit the company to be more competitive in its industry, thus helping to safeguard jobs.
>
> Signed .... Date.....................
>
> (Managing Director)

Scope of rules

Rules should cover aspects such as the following:

- "housekeeping";
- use of machinery;

¶704

- use of protective equipment;
- maintenance procedures;
- accident reporting;
- first aid attention;
- fire prevention;
- electrical restrictions;
- use of tools and equipment;
- authorised entry and usage (certificates, licences);
- smoking, alcohol and drugs;
- prohibition of horseplay and misuse of equipment; and
- any other rules applicable to the organisation's work.

Example

The following example shows a typical layout and content of a set of basic health and safety rules. Any set of rules will need to meet the needs of an organisation or work area.

In addition to basic safety rules, the organisation may make use of work method statements. These are documents that set out exactly how particular tasks should be performed. These will discourage the taking of risky shortcuts. They should be written in plain English; workers should be trained in their application and supervision should be adequate to ensure they are adhered to.

Example
Basic health and safety rules

These basic health and safety rules apply equally to every person working within this organisation.

Any breach of one of these rules may result in serious injury to one or several people. Therefore, each supervisor has the duty and authority to take appropriate disciplinary action after any deliberate violation of any of these rules.

1. **Know and observe departmental health and safety rules**

Each department or section has its own specific operating rules and procedures in addition to these basic health and safety rules. Each employee must be aware of, and follow at all times, these rules and procedures when performing the job.

2. **Observe** *all* warning signs and danger tags. Keep *all* safety guards in place

A warning sign, danger tag or safety guard is put there to ensure your protection from recognised hazards. This rule is necessary to protect everyone.

3. **Use safety glasses and other protective equipment where required**

Employees owe it to themselves, their families and the company to use and properly care for the personal protective equipment provided for on-the-job use. Any faults in the condition of any of the equipment should be reported to supervisors.

4. **Report** *all* injuries or potential hazards immediately

A first aid station is provided for treatment of injuries in all work sections. Prompt treatment for even small injuries is the best way to prevent painful and costly complications. All injuries must be reported as soon as possible to the first aid station and details of accidents reported in the accident book. The supervisor must also be notified.

5. **No smoking on the job**

Because of several highly flammable chemicals used in work processes, flames and sparks must be carefully controlled to avoid fire or explosion. Smoking is permissible only in designated areas.

6. **No alcohol and drugs on the job.**

People under the influence of alcohol or inappropriate drugs can make errors in the performance of their work, as well as presenting a risk of accidents or injuries to themselves or others. Your work requires the maximum degree of alertness and good judgement. If your doctor prescribes drugs to treat a medical condition, you will need to advise your doctor of your work environment so that he or she can determine whether it is safe to continue working whilst taking that medication.

7. **No horseplay**

Serious injuries and incidents can result from horseplay or practical jokes.

Signed.....................Date......................
(Managing Director)

¶704

¶705 Recruitment and selection of employees

The job specification for each position should contain the particular skills, knowledge and abilities that are required by a person to be able to successfully undertake the tasks involved in the job in a safe manner. There may also be particular statute or award provisions applying to junior employees, for example.

Examples of requirements could include physical fitness, eyesight, mental alertness, height and weight (only where relevant), education, language, possession of particular skills and qualifications, and general health (see medical examinations at ¶720). These requirements should only be included where they can reasonably be considered to be essential for a position. Where this is not the case, including such a requirement in the criteria for a position could amount to a breach of equal opportunity legislation.

Job descriptions should include more general safety statements. For example, a job description could include the following as a required duty:

"Comply with safe working procedures. Monitor and maintain a safe working environment and report any health or safety hazards to area supervisor."

The job description for a storeman could include the following:

"The worker will be expected to utilise safe manual handling procedures in accordance with training."

Statements such as these, while being general, will emphasise the importance the organisation places on health and safety. Similar statements can be placed in advertisements for positions.

Performance criteria for supervisors and line managers should also incorporate measurable performance indicators relating to health and safety objectives and targets. These will provide management with feedback on what is happening. They can be outcome-based, reflecting system or operational performance (eg. rate of injury), or they can be input based (eg. number of inspections conducted, instruction and training provided or audits performed).

HEALTH AND SAFETY TRAINING

¶706 Why train?

As stated in Chapter 2, one of the requirements of both common law and legislation is that management must provide competent staff. A disproportionately high number of serious injuries are suffered by either new employees or those commencing a new or different work activity and sometimes after returning to work after leave.

The basic aim of health and safety training is to impress the principles of health promotion, risk management and safe behaviour on employees in such a way that they will apply these principles to their work. The training required will vary according to position held and type of work performed.

A structured health and safety training system should reduce the number of injuries at a workplace, and should be measured against that criterion. Associated benefits will be an increase in productivity and a reduction in the costs associated with injury and disease. Further, a formalised training system emphasises an organisation's commitment to a safe workplace; it creates a "culture of safety" in the organisation. It also facilitates the assessment of an employee's competence as required under common and statute law.

¶707 Induction training

Induction training is usually the first introduction to an organisation. It is a combination of a formal introductory session and basic on-the-job training that can be conducted by a supervisor. Parts of a program should also be used when an employee is transferred into a section/department from another within the organisation, as they are "new" to that area.

Induction training should include:

- general instruction on the workplace and work performed;
- notification of the organisation's health and safety policy and any health and safety programs in place;
- information on risk management and expectations of the employee in terms of participation in the process;
- advice of health and safety rules (see ¶704);
- issue, and advice on use of protective clothing and equipment (see Chapter 9);
- advice on particular hazards at the workplace;
- response in the event of fire or other emergency;

- instruction in necessary skills, such as manual handling, machine operation;
- advice on first aid facilities and personnel;
- procedure for workers' compensation claims; and
- advice on, and identification of, the health and safety representative and/or safety committee.

¶708 Continuing training

Effective training is based on needs rather than mere routine. It may be appropriate to target a group of employees who work in a particularly dangerous area, or where statistics reveal a problem, or where a new process is introduced.

The content of any training should be documented and periodically reviewed. Records should be kept of who has received training, and of the content of that training.

Note that there are specific outside courses available on particular subjects such as manual handling, fire safety and hearing conservation. Training organisations can be contracted to provide such courses, either on-site or away from the workplace. Another option is to contract a training organisation to design an on-site course specifically to suit the needs of a particular organisation. In some areas, such as crane driving and fork-lift driving, legislation requires operators to hold an appropriate certificate.

Preparing training

The first step in preparing any training program is to undertake a training needs analysis to determine if the problem can be resolved by training. If training can assist, it is then necessary to define the outcomes/objectives required. These outcomes must be realistic, and at the end of the training program it must be possible to determine whether they have been met. The desired outcomes should be communicated to the trainees at the commencement of training.

Once the outcomes have been set a training program that will achieve those objectives can be designed. Classroom training or on-the-job training are two alternatives. Other approaches include a supervised project undertaken at work, observation of other employees performing the task (preferably followed by a "try out" of the task by the trainee), or a self-directed group exercise.

Different levels of employees

Different training needs for different levels of employees are set out below.

Implementing a Program 133

Employees

All employees need to be sufficiently informed about safe methods of performing the work in which they are engaged. They should understand their role within the organisation's overall health and safety policy, and in the risk management process.

The actual level of training required will depend on the danger and complexity of the tasks performed. An indication of the need for training can be obtained from the safety statistics for each group of employees.

Non English speaking employees

The fact that communication is an acquired skill even when instructor and trainee normally speak the same language underlines the necessity for special measures when they do not. Where there are employees who do not speak or understand English well, safety problems can arise if they do not understand instructions and are afraid to admit it. Some words can have similar sounds but totally different meanings.

Measures to remedy this situation include appointment of bilingual employees as interpreters, and written instructions in other languages. Both methods are appropriate, although both have potential problems. Interpreters may tend to develop into unofficial supervisors although they are not suitable in other ways. Written instructions fail when ethnic employees are illiterate or only partially literate in their own language. The overriding concern is to ensure that all employees understand the safety instructions that are given. Demonstrations and the use of pictures or videos can assist in imparting the safety message where there are language problems.

Supervisors should make allowances for the difficulties (and hence the added risks) encountered by workers from a non-English-speaking background. Above all they should ensure that instructions are given in the simplest terms and in the clearest way possible.

Classes for workers to learn English can be arranged at work locations. They are usually arranged partly in work time and partly in employees' time, with the human resources department co-ordinating the classes. Enquiries should be directed to the federal Department of Immigration and Ethnic Affairs. In NSW the Adult Migrant English Service provides English language classes in the workplace. Note that some awards, most notably in Western Australia, contain a provision requiring an employer to allow migrant employees to attend English language classes during work time without loss of pay; and in Victoria a code of practice requires the dissemination of information in appropriate languages.

¶708

Health and safety representatives/committee members

Training is crucial to enable health and safety representatives or committee members to carry out their roles effectively. This is discussed at ¶507 and ¶508. Note that the legislation generally provides that an employer must grant a representative (or committee member in NSW) time off without loss of pay in order to attend accredited training courses.

Supervisors

The supervisor's control over the workplace, its people and resources necessitates an understanding of risk management and accident causation. The supervisor's role is to manage these. Health and safety must be accepted as part of the supervisor's job in the same way as all other duties and activities.

For these reasons, there is likely to be more emphasis on formal health and safety training courses, run by either the organisation's training section or an outside organisation such as a government department (see Chapter 10). It is also important to remember that supervisors will be responsible for a large part of the health and safety training of their subordinates. Therefore they need skills that include training techniques as well as health and safety management, report writing, investigation methods and motivational techniques. This may involve their attendance at courses.

Supervisors may also find it beneficial to belong to some outside health and safety organisation to keep abreast of latest developments. Finally, they may benefit from private research and should be on the circulation list of health and safety journals and related publications received in the organisation.

Senior managers

The approach for senior managers will be similar to that for supervisors, particularly in smaller organisations. Health and safety rules and policies should apply to everyone from the highest level of management down. Therefore, health and safety rules and procedures should be included in the induction and training of senior staff.

Managers are likely to have less direct contact with on-the-job matters than supervisors, so their training emphasis will rest to a greater extent upon health and safety reports/statistics and monitoring. In many cases, it will help to provide information and results from a viewpoint that emphasises costs and profits. Senior managers should make regular visits through their entire area of control and show interest in the health and safety function by discussing it with supervisors.

¶708

COMMUNICATING HEALTH AND SAFETY

¶709 Creating a high profile

Merely training employees in healthy and safe working practices will often not suffice. In many cases it will be necessary to provide forms of motivation and publicity that encourage them to continue to take an active interest in self-preservation and the health and safety of others.

For supervisors and managers, one effective approach could be to include the section's health and safety record as a part of the organisation's performance assessment system.

The methods used will need to create and reinforce an atmosphere of healthy and safe behaviour among employees and point out the benefits for employees, as well as the organisation, of adhering to working practices that promote this. The closure of a business due to excess costs caused by a high injury rate would lead to jobs loss.

The topics already discussed in this chapter (the publication of a health and safety policy, the inclusion of safety statements in job specifications and the training of employees) will all promote safety as an important issue within the organisation. The appointment of health and safety representatives and/or committee members from among the employees also has a positive effect on the promotion of health and safety. Other common promotional techniques are listed below and discussed in the following paragraphs.

Health and safety promotion approaches

- Posters
- Incentive schemes and competitions
- Printed matter
- Videos

¶710 Posters

Posters and information sheets may be obtained from a wide range of sources, such as government departments, employer/employee associations and trade houses, as well as private organisations (see Chapter 12). The posters chosen should either be directly applicable and related to the particular work area or else be of general appeal or message. They must not be counter-productive, such as the phrase "Make a strike for safety" might imply.

Posters can overcome language problems through the use of illustrations and symbols.

To maintain attention, posters should be kept on a special display board (not cluttered up with other notices) and changed at frequent intervals. Posters do not constitute a health and safety program in themselves, but have greater value when integrated with an overall planned program.

¶711 Printed matter

A wide variety of publications on safety and health matters is available. They range from simple instructional leaflets on particular topics (such as lifting, fire-drill, electrical safety, personal fitness) through general items (such as checklists and guides to legislation and sources of further information) to more detailed reports, books, etc.

Many publications are designed for use by management and supervisors rather than shop floor employees, so it is important to check them for suitability before distributing them. Fact-sheets and smaller instructional leaflets can be suitable for general distribution and should be used at every opportunity.

A vast array of publications of this type can be found on the internet, at the websites of the National Occupational Health and Safety Commission and State and Territory OHS authorities (see Chapter 12).

Note again the advantages of issuing printed matter in several languages if the work-force is multicultural. It is also possible to make use of the organisation's existing publications such as the company's in-house newsletter.

Merely distributing pamphlets without any other action may not be very effective. It is better to draw attention to the information by holding short meetings to discuss the subject of the publications, reinforce the high priority the organisation puts on health and safety, and encourage workers to read them.

Safety/health library

Many organisations may benefit from setting up a small "library" of health and safety publications. This should include copies of relevant legislation, regulations, awards and codes of practice, as well as detailed information on specific topics. Information on health problems such as stress or alcohol abuse, for example, could be useful when counselling employees. Cost of publications should be included in the health and safety budget.

Access to a library is also a means of providing supervisors with more information. It can be complemented by a circulation list to ensure staff

Implementing a Program

receive current journals. A library can also assist the preparation of health and safety training sessions.

¶712 Videos

Video is an excellent way of conveying health and safety messages, and there are many available on a wide variety of health and safety topics. They can be powerfully exploited for the purpose of demonstration, and many are supplied as training packages with accompanying manuals for trainers and participants.

To have the maximum impact, videos on occupational health and safety should convey a sense of reality. Local productions (ie. involving Australians) may have more credibility than equally informative videos from overseas.

When selecting videos, it is worthwhile to consider that strong emotional appeals or grisly accident scenes that intend to "shock" viewers sometimes prove ineffective. The viewers may regard these tactics as too far removed from the reality of their own lives, and thus fail to perceive the situations as being applicable to themselves. Identification can be better achieved by following a video with a slide presentation or discussion illustrating local incidents and injuries or structuring a discussion to make it relevant to the particular workplace.

Some information on possible sources of health and safety videos is provided at Chapter 12.

Further references:

The *Journal of Occupational Health and Safety — Australia and New Zealand* (CCH Australia Limited) carries reviews of films and videos in most issues.

¶713 Incentive schemes and competitions

The use of health and safety incentive schemes or competitions is a controversial approach. While they appear to work effectively in many organisations, the logic behind them is questionable. In addition, they may eventually become ingrained as a form of "over award" payment or benefit, and no longer directly associated with health and safety.

The common form of scheme is to present bonuses, gifts or prizes to employees or employee groups for achieving certain "target" levels of accident-free or injury-free working hours. In the case of competitions, different groups of employees, such as separate departments or plants of the same company, compete for the reward offered. Note, however, that for income tax purposes the value of such gifts may be regarded as income from work and therefore needs to be declared as taxable income. From an employer's point of view, awards genuinely related to safety achievements, up to a value of $200 per annum per employee, will be exempt from Fringe Benefits Tax.

The basic advantages and disadvantages of incentive schemes and competitions are set out below.

	HEALTH AND SAFETY INCENTIVE SCHEMES/COMPETITIONS	
For		*Against*
1. Helps develop a collective concern to prevent accidents. 2. Employees more conscious of the need to spot and control hazards. 3. Motivation through linking safety directly to income.		1. May encourage people not to report accidents or injuries when they occur, for financial reasons. This could lead to a recurrence of the accident/injury. 2. Arguments over the definition of "disease", "accident" and "injury" may occur. 3. Administrative problems — such as equating sections performing different types of work on a competitive or handicap basis. 4. The questionable logic of rewarding employees for not injuring themselves. 5. The "carrot-and-stick" motivation approach inherent in these schemes may fail if one of the two is removed.

¶714 Occupational health service

Large companies may choose to establish an in-house occupational health service to help protect workers from occupational injury and disease. Such a service is usually staffed by occupational health nurses and/or occupational physicians. Its functions include the promotion of a healthy work environment and general health supervision, monitoring the work environment, treatment and rehabilitation, education and counselling, supervision of first aid, record keeping, research and liaison.

Location and facilities

The location, layout and facilities of an occupational health service should be planned either in conjunction with existing occupational health and safety staff or with the relevant State occupational health and safety authorities. The planning and equipping of a health centre should comply with any existing legal requirements or established codes of practice. The size of the particular business or organisation, its type of operation and, therefore, the kinds of work hazards present should also be taken into account. It should be noted that various casual, on-site services (such as audiometric testing) are now available commercially.

Smaller organisations

Smaller organisations may not be able to justify the expense of an on-site health service, and may wish to contract an outside service to provide some of the same functions. This may involve an initial familiarisation with the workplace, regular visits by the health service to monitor the workplace and perform health checks on workers, as well as workers and management having access to the health service at all other times. Because workers may need to attend the health service following an accident, it is desirable to engage a health service near the workplace.

REHABILITATION

¶715 What is occupational rehabilitation?

The majority of employees who are ill or injured simply return to work after basic first aid and medical treatment. Occupational rehabilitation addresses the needs of the remaining minority who are more seriously affected.

Rehabilitation — sometimes referred to as injury management — means the restoration of the disabled to the fullest physical, social, vocational and economic usefulness of which they are capable. Organisations are expected to do everything reasonably possible to safely return an employee to work. Once medical needs are attended to, a graduated return to work can be devised the workplace may be modified and job tasks redesigned, or the worker could be assisted to retrain for another position. As well as looking at vocational aspects, effective rehabilitation should address any psychological or social problems an injured worker may be having.

Benefits of rehabilitation

There has been an increased emphasis on rehabilitation (often referred to as injury management) in recent years as employers come to realise the benefits obtainable. Effective rehabilitation will mean that employees are able to return to work sooner, with the consequential savings in workers' compensation costs. As it is the minority of long-term injuries that make up the large majority of workers' compensation costs, cases of organisations halving their workers' compensation costs have been reported. In addition there are intangible benefits, such as increased staff morale.

Of course the main beneficiary from rehabilitation is the injured worker. Long-term loss of the ability to work usually leads to a loss of status and morale, and sometimes financial difficulties. The best way to address these problems is to engineer a return to work as soon as it is safely possible.

Legislative encouragement of rehabilitation

In recent years the workers' compensation legislation in most Australian States has been amended to place increased emphasis on rehabilitation. In most States employers are now required to develop rehabilitation programs and/or to provide suitable duties for injured employees.

The requirements in each State are mentioned at ¶718.

Further references:

Workers Compensation in Australia, Industry Commission Report, 1994.

Guidance Note for Best Practice Rehabilitation Management of Occupational Injuries and Disease, National Occupational Health and Safety Commission, 1995.

Uniform Guidelines for the Accreditation of Rehabilitation Providers, National Occupational Health and Safety Commission, 1995.

Rehabilitation Guidelines: Guidelines for Workplace-based Occupational Rehabilitation Programs for Large and Medium Sized Businesses, NSW WorkCover Authority, 1994.

Workplace Rehabilitation Manual, CCH Australia Limited, 1990.

¶716 A rehabilitation program

Medium and large organisations will find it cost effective to introduce a rehabilitation program that sets out the procedures to be followed in the rehabilitation of individual employees (note that this is a statutory requirement in New South Wales, Victoria and the Australian Capital Territory — see ¶718). The program can comprise a rehabilitation policy and detailed procedures to implement that policy.

A rehabilitation policy

A rehabilitation policy can be used to introduce rehabilitation into an organisation. The policy will demonstrate an organisation's commitment to rehabilitation, publicise the rehabilitation program, and set out who will manage the program and the parameters within which the program will operate.

A sample rehabilitation policy is provided on the following page.

Rehabilitation procedures

The company should ensure that there are clear written procedures to follow for every person who sustains a work-related injury or illness. These procedures should include:

- injury reporting;
- first aid;
- arranging for medical or other treatment;
- completion of necessary forms (accident records, workers' compensation);
- liaising with medical professionals and insurers;
- maintaining contact with the injured employee;
- determining suitable duties for the rehabilitating employee;
- the process to be followed when a person returns to work; and
- monitoring and upgrading each return to work.

These procedures, together with the policy discussed above, will form the organisation's rehabilitation program.

Who should coordinate rehabilitation?

Organisations may wish to appoint a rehabilitation co-ordinator to manage the rehabilitation of injured employees (note that this is a statutory requirement in New South Wales see ¶718). This person will be responsible for liaising with the various parties involved in the rehabilitation process: the injured worker, the worker's own doctor, the company nurse/doctor, the rehabilitation service providers, and the supervisor in the area where suitable duties are chosen. The rehabilitation co-ordinator will also be responsible for monitoring the progress of each rehabilitation case, and preparing reports on the outcome of each case.

Where the organisation has a company health service, it may be appropriate to appoint someone from within that service to act as the rehabilitation co-ordinator. In other instances the company nurse, the safety officer or someone from within the human resources department could be chosen.

¶716

Rehabilitation policy

The company is committed to preventing illness and injuries at the workplace by providing a safe and healthy working environment for all our people. It is recognised that injury or illness may still occur and therefore all incidents will be reviewed and steps will be taken to prevent recurrence.

The company believes that occupational rehabilitation is of benefit to everyone and should commence as soon as possible following injury or illness. Furthermore, no person being rehabilitated will suffer financial loss or prejudice in any way.

Early reporting of injury and illness is encouraged.

The company will ensure access to good first aid and high quality medical care. Accurate medical diagnosis and assessment will be followed by early intervention from a rehabilitation service provider if necessary.

Every effort will be made to assist people in an early, safe return to meaningful and productive work in consultation with their treating practitioners.

Suitable duties will be provided by the company; where this is not possible, early referral to a rehabilitation unit will be facilitated.

A graduated return-to-work program consistent with medical advice will be followed. Each person will be given a written return-to-work program. The rehabilitation co-ordinator will assist in this process by providing the necessary link between treating practitioners, rehabilitation service providers and the workplace.

The company has consulted with employee and union representatives; there will be no inter-union disputes (demarcation disputes) arising because some people may have to be rehabilitated with alternative duties outside their usual job classification. However, adequate training for such alternative duties will be given to ensure that safe working practices are followed.

The company's rehabilitation co-ordinator is

....

The company's chosen rehabilitation provider(s) is/are:

....

(name and address)

....

The injured employee retains the right to choose his/her own treating doctor and to choose an alternative rehabilitation provider.

All people on site have an important role to play to ensure the best possible outcome for their injured colleagues; successful occupational rehabilitation requires everyone's involvement and commitment.

Employee Representative:

....

Managing Director:

....

¶716

¶717 Features of occupational rehabilitation

Some of the important features of occupational rehabilitation are outlined below.

Rehabilitation providers

Occasionally, in cases of serious injury and long periods away from work, an injured worker will require the services of a multi-disciplinary rehabilitation team. Such teams may comprise various health professionals (doctors, nurses, physiotherapists, occupational therapists, etc) as well as counsellors (psychologists, social workers, career counsellors, etc). An organisation should select one (or several) rehabilitation provider(s) in the local area capable of meeting the organisation's needs. Note that in some States certain rehabilitation services have been declared as "accredited rehabilitation providers" (see ¶718). Lists of accredited providers can be obtained from the relevant authority.

Rehabilitation providers may be private organisations, units attached to hospitals, or the Commonwealth Rehabilitation Service.

In simpler cases an injured worker may only require the services of one particular health professional, such as a physiotherapist or a chiropractor.

Early intervention

Early reporting of symptoms can allow a problem to be identified and rectified before it becomes chronic. For example, a word processor operator may report sore wrists or a stiff neck as soon as those symptoms are recognised. Modification of the work station or work processes can avoid serious disruption to the employee's life and to the employer.

Early assessment of the need for rehabilitation following an accident is also important. As soon as possible the injured worker should aim towards a return to work, rather than becoming accepting of the "sick role".

Participation of the injured worker in decision making

Too often injured workers are made to feel that they have no control over decisions concerning their health. An injured worker who is part of the decision-making process is more likely to be committed to a successful outcome.

Injured employees must retain the right to choose whether to participate in rehabilitation treatment, and the right to choose their doctor and rehabilitation provider. The legislation in some States, however, provides disincentives (a reduction in benefits) where people refuse to participate in a rehabilitation scheme.

Suitable duties

Suitable duties are the cornerstone of any workplace rehabilitation program. They allow the employee to return to the workplace, increasing fitness and reducing any depression or "compensation syndrome". For the employer, suitable duties can greatly reduce the working days lost due to injury.

Suitable duties are tasks designed specifically for an injured worker in order to allow an early return to work. The work must be carefully matched to an employee's capabilities so that it facilitates recovery while not risking further injury.

The duties provided may be some of the employee's previous tasks, or they may be the previous tasks performed at a slower speed or less weight, or they may be completely different tasks. Any planning of suitable duties should be done in consultation with the injured worker, his/her doctor and the employees and supervisors who will be directly affected.

It is important to monitor and upgrade a program of suitable duties, to ensure the worker is progressively moving towards full recovery. Gradually increasing the time spent at work is one common way of achieving this.

Problems can arise when suitable duties are provided in an area other than the injured worker's usual area of work. The injured worker may be working in an area covered by another union, and may be on a different rate of pay from other employees performing the same work. These issues should be addressed in the organisation's rehabilitation policy. It is important to ensure that a rehabilitation program does not create any risks/hazards for other employees by making changes to their jobs to accommodate an injured employee.

In some States there is a legal requirement to provide suitable duties see ¶718.

¶718 Rehabilitation systems — State by State

The following provides a brief description of the government regulation of rehabilitation in each State. Further information can be provided by the relevant authority in each State.

New South Wales

The *Workplace Injury Management and Workers Compensation Act 1998* introduced the concept of injury management — a "coordinated program of treatment, claims management, rehabilitation and employment practices to achieve a safe, timely and durable return to work for injured workers" — as a cornerstone in the management of workplace injuries in New South Wales.

Implementing a Program

The key features of the system require insurers to develop injury management programs governing their procedures for managing workers compensation cases, and employers to develop return to work programs that comply with their insurers' injury management programs. Employers' return to work programs must address the commitments and procedures set out in the WorkCover NSW *Guidelines for Workplace Return to Work Programs*, that apply in total to larger employers, ie. those with a basic tariff premium of more than $50,000, or self-insurers, or employers with more than 20 employees, who are insured with a specialised insurer. All other employers (who do not meet any of the above criteria) may use a WorkCover NSW standard workplace return to work program.

The system also requires employers to notify their insurer of significant injuries (those requiring more than seven days away from normal duties) within 48 hours. Workers with significant injuries must nominate a doctor to oversee their injury management. The insurer, in consultation with the employer and the nominated treating doctor, will develop an injury management plan for individual workers with significant injuries. Employers are expected to make suitable duties available to injured workers.

Victoria

Rehabilitation in Victoria is administered by the Victorian WorkCover Authority.

Certain employers (those with a rateable remuneration of over $1 million annually) are required to establish and maintain an occupational rehabilitation program. If a worker is incapacitated for 20 or more days, the employer must prepare a return-to-work plan for that worker, and nominate a return-to-work coordinator.

Employers must keep jobs open for a period of 12 months to provide suitable employment for injured workers. Workers may have their weekly benefits stopped if they refuse to make reasonable efforts to return to work, refuse to accept a suitable job offer, refuse to undergo training or refuse an assessment as to rehabilitation prospects.

Queensland

Queensland employers with 30 or more workers at a workplace must have a workplace rehabilitation policy and procedures in place, review the policy at least every three years (and provide WorkCover Queensland with written evidence, in the approved form, of the review's completion), and have the policy and procedures approved by WorkCover.

Employers must assist in, or provide injured workers with, rehabilitation, which must be of a "suitable standard". A rehabilitation plan

¶718

must be developed for each worker who is incapacitated from work. Employers with 30 or more workers must also appoint a fully trained rehabilitation coordinator, who must be accredited by WorkCover. Employers must provide injured workers with suitable duties, and workers cannot be dismissed for three months after the injury for reasons relating to the injury.

South Australia

The WorkCover Corporation is responsible for the rehabilitation of injured employees. The Corporation is required to establish or provide rehabilitation programs, and can appoint rehabilitation advisers.

Employers must provide suitable employment for their injured employees, unless the employer can show that this is not reasonably practical. Employers with 10 or more employees must keep an injured employee's position open indefinitely, and those with less than 10 employees must keep the position open for 12 months.

Western Australia

The relevant legislation in Western Australia "encourages" employer-based rehabilitation programs. The Workers Compensation and Rehabilitation Commission has the authority to require an employer to take reasonable steps to facilitate the rehabilitation of a worker.

Where feasible, employers must provide rehabilitated workers with their original job prior to injury. If that position is no longer available, or if the worker can no longer perform the job, the employer must offer a similar position that the worker is qualified for, capable of doing, and that commands comparable status and pay.

Employers must keep an injured employee's job open for 12 months and must take reasonable steps to rehabilitate the employee.

Tasmania

A rehabilitation policy must be prepared by employers of more than 20 workers and it must be displayed at the workplace.

If a worker is totally or partially incapacitated for more than 14 days, the employer must prepare a return to work plan in consultation with the worker. This must be done within five days of the end of the 14-day period of incapacitation. Where feasible, employers must offer suitable alternative duties that the worker could reasonably be expected to perform. In addition, employers must keep an injured worker's job open for 12 months unless the reason for the original employment no longer exists, or it is not feasible to make that employment available to the worker.

¶718

Implementing a Program

Australian Capital Territory

The employer must provide the worker with occupational rehabilitation at the employer's expense. Employers must develop an occupational rehabilitation policy and review it from time to time. A rehabilitation coordinator must be appointed by the employer. The Chief Minister's Department of the ACT administers the legislation.

Northern Territory

Northern Territory employers must bear the reasonable costs incurred for a long-term incapacitated employee's rehabilitation training. This includes a workplace-based return to work program and workplace modification. As far as possible, employers must provide injured workers with, or assist them in finding, suitable employment, and participate in efforts to retrain workers.

¶719 Safety and health off the job

The best approach to creating improved safety and health behaviour amongst employees is to attempt to make it encompass all aspects of their lives (that is, safety at home, on the road, in the water, etc). The aim is to make healthy and safe behaviour "automatic" so that employees observe these precautions as a matter of course and do not have to be constantly reminded or encouraged.

This approach has benefits both ways. The organisation benefits through lower rates of absenteeism, forced retirement of key personnel, labour turnover, and so on. For the employee, there are extra benefits because many of the health and safety aspects learnt at work can be used outside work, such as proper manual handling techniques, fire control, first aid and resuscitation.

Some employers provide assistance to employees to improve their general health They know that the improvement in health will benefit the company as well as the individual. On-site gymnasiums or sporting facilities are perhaps the most common examples of assistance. There may be some risks associated with on-site facilities if they are not supervised by appropriately trained personnel. The facilities are paid for by the employer, and employees are encouraged to use them in their own time. Other alternatives include paying membership fees to a gymnasium, and sponsoring sporting teams made up of employees.

Another aspect of employee health can be addressed through employee assistance programs. These are counselling programs offered to employees with problems such as drug and alcohol abuse, marital problems and financial worries. The rationale for an employer providing such programs is the premise that personal problems will affect an employee's work performance. Employee assistance programs are discussed in more detail at Chapter 11.

¶720 Medical examinations

An organisation may undertake medical examinations of its employees for a variety of reasons:

1. *Pre-placement*: to determine an applicant's suitability for a particular job. For instance, has the person sufficient physical strength if the job requires heavy work; is eyesight sufficiently keen to meet the requirements of assembly of small parts; checks for colour-blindness may be necessary in some industries such as electrical trades; is the person prone to allergies or bronchial trouble if working in particularly dry, moist or dusty atmospheres? The particular requirements will obviously vary from industry to industry, and are designed to protect both the employer and the employee.

The examination, if part of the selection procedure, should be held near the end of the selection process, so that only applicants who appear otherwise suitable are tested. This has the advantages of saving time and money, and placing the test as near to the person's entry into your organisation as possible. The examination is a record of a person's fitness at the time of hiring and further examinations will show deviations that could be of assistance in workers' compensation claims or common law suits. Pre-placement examinations should also be used where employees are transferred within an organisation.

Note that the aim of a pre-placement examination should not be to exclude persons who fail to measure up to a rigid health standard but to check their suitability for the type of work involved. It may be that people with certain medical disabilities would still be suitable for other jobs within the organisation.

In recent years there has been some controversy over pre-placement testing to screen for alcohol or drug use. This is becoming more common in North America, but is not yet widespread in Australia. While there may be a case for screening in certain jobs (airline pilot, train and coach drivers), cost and privacy considerations make it less acceptable in other situations. The provision of employee assistance programs (see ¶1203) may be a better alternative.

2. *Periodical examinations of employees*: most commonly on an annual basis. These examinations can be either of all employees or of subgroups (such as executives). The purpose of these examinations can include both general health (blood tests, electro-cardiograms, radiography) or tests for specific occupational hazards (industrial deafness, exposure to potentially harmful

Implementing a Program 149

substances and work processes, including zoonoses [diseases acquired from animals], silicosis and asbestosis).

3. *Entry to superannuation and disability schemes.*

4. *Workers' compensation*: to determine fitness to resume usual work activity, or for placement into alternative work situations that would not affect the type of injury received but provide a meaningful work task.

5. *Testing of employees for AIDS*: in practice, there are very few industries where there is any possibility of transmission of the AIDS virus at the workplace. With this in mind there seems to be little justification for testing for HIV infection at the workplace. In any event, testing should not be performed without an employee's consent, and any test results should be treated in the strictest confidence. An employer who dismissed or transferred an employee because he/she was HIV positive may be guilty of unlawful discrimination and be liable for damages. Further information can be found in the book *AIDS and the Workplace: A Practical Approach*, by Dr G Tillett, CCH Australia Limited, 1989.

> Medical examinations should not be perceived by management or employees as a means of penalising employees for any health problems they may have developed, but as an effort to ensure the continued good health and safety of all employees.

A major advantage of regular examinations is that any problems, particularly ones partially or wholly due to factors at the workplace, may be identified and corrected before they become serious. Statistical data from examinations could also be used in some cases as a means of approaching and reducing the causes of health problems.

If the organisation lacks the facilities to conduct or arrange medical examinations of employees, it may still be a good idea to encourage employees to voluntarily arrange their own medical examinations at regular intervals.

Confidentiality issues

The collection and use of medical information on employees is a very sensitive issue.

Employers should ensure that medical reports are kept confidential, and that medical records are not kept any longer than a legitimate use for them exists. For example, the NSW Privacy Committee recommends that employers should destroy medical reports of every unsuccessful applicant upon request of the applicant or within two years after application for

¶720

employment is made, unless exceptional circumstances exist or the applicant requests that the report be kept for a longer period.

The employer is not the appropriate person to discuss the contents of a medical report with an applicant. If details of medical reports are requested by an applicant, the person should be advised to contact the examining doctor or qualified staff. If the employer's consent is required for the release of details, it should not be denied.

Note that a neutral stance by the doctor will be essential. The doctor must never be perceived as catering for the interests of management. A rigid code of ethics for communication of information will need to be worked out and adhered to.

If medical records are to be retained by employers, access to them must be strictly controlled. It would seem better, however, for any detailed records to be kept by the doctor and general advice given by the latter to employers.

¶720

Chapter 8

Evaluating Health and Safety Performance

The importance of evaluation	¶801
Choosing measurable objectives	¶802
Assessing results against original objectives	¶803
Employees' right to know	¶804
Health and safety auditing	
Auditing to control risks	¶805
Conducting an audit	¶806
Scope of the audit	¶807

¶801 The importance of evaluation

Successful health and safety performance cannot be assumed merely because an organisation has implemented a particular program. Like every other aspect of the organisation's performance, each health and safety program must be evaluated by management in order to determine whether it has been successful and whether any improvements can be made. Without the necessary resources devoted to evaluation, the organisation is committed to "blind acceptance" of the program originally chosen.

Evaluating health and safety performance has the following elements:

- choosing measurable objectives;
- measuring what has happened;
- comparing these measurements against the original objectives; and
- deciding on new or revised plans.

¶801

Methods of measurement

Methods of measuring the outcome of a safety program will depend on the program. For example, a hearing conservation program can be assessed by taking measurements of noise levels and by audiometric testing of employees. Further examples are given at ¶803. Safety auditing, a form of ongoing health and safety measurement, is discussed at ¶805 et seq.

¶802 Choosing measurable objectives

The objectives of each health and safety program will depend on the program and the workplace. What is important is that the objectives chosen are measurable, rather than vague statements. Performance objectives should be based upon the "SMARTA" principle, that is, they should be Specific, Measurable, Achievable, Relevant, given a Time-frame, and Agreed.

For example, "a better place to work" or "cleaner air in the laboratory" are not measurable objectives. Better choices would be "a 30% reduction in labour turnover within 12 months" or "reducing the concentration of compounds X, Y and Z in the laboratory atmosphere to below the National Occupational Health and Safety Commission's published exposure limits". Further examples of objectives are given at ¶803.

Where the program is one of ongoing improvement, a time frame should form part of the objective. In some cases improved productivity, or increased safety with no loss of productivity, can form part of the objective.

¶803 Assessing results against original objectives

The reason for setting measurable objectives is to enable the outcome of a program to be assessed, and to make changes that are shown to be necessary.

Examples

A hazardous substances safety program

The objectives of a hazardous substances safety program were agreed to be:

- catalogue and prepare MSDS (Material Safety Data Sheets — ¶914) for all chemicals and other hazardous substances at the workplace;
- adequately label all chemical storages at the workplace;
- ensure that showers and eye-baths are available at all locations where they may be needed; and
- reduce the atmospheric level of specified chemicals to below agreed levels.

¶802

Six months was set as the target time for achieving these objectives.

At the end of six months a survey of all chemicals and other hazardous substances in the workplace had been carried out, and all chemical storages had been labelled in accordance with the National Health and Safety Commission's guidelines. Showers and eye-baths had been placed within easy access of all major chemical usage areas at the workplace.

Measurements of the specified chemicals showed that atmospheric levels had in fact been reduced to below internationally accepted exposure standards. However, MSDS had not been compiled for some of the chemicals that were catalogued. The reason for this was a delay in getting information from certain chemical supply companies. It was decided to put pressure on the companies to supply the necessary information, and to investigate other suppliers if information was not forthcoming within two months.

A program to restrict smoking at the workplace

The agreed objective of a program to restrict smoking at the workplace was:

- smoking is only to be allowed in certain designated areas that should be appropriately labelled.

A period of 12 months was allowed for the implementation of the program. Attempts were to be made to meet all genuine grievances of smokers and non-smokers.

At the end of 12 months smoking areas had been selected and labelled. While compliance was good, the occasional breach was justified by smokers on the grounds that the designated areas were often too crowded and too smoky. As it was not possible to provide additional smoking areas, it was decided to amend the program by placing chairs and ashtrays outside certain entrances to the building in an attempt to encourage smokers to use these areas during breaks. Mechanical ventilation of the smoking areas was also provided.

¶804   Employees' right to know

Employees at a workplace are entitled to know the results of any evaluation of a health and safety program in which they have participated. Management and employees jointly working towards improved health and safety suggests that all parties know of any programs that are introduced and the results of those programs. Note that legislation generally provides health and safety representatives (or committee members in NSW and the Northern Territory) with the right to all safety information concerning the workplace — see ¶204.

It may be worthwhile for management to call a meeting of all employees involved in a particular program after an assessment of that program has been completed. The results of the assessment can be presented, and options for amending the program could be discussed. Where a large number of employees is involved (such as a program that affects the whole workplace), this process could be carried out through the health and safety representatives and/or committee. All staff should be informed by circular or a message on safety notice boards.

HEALTH AND SAFETY AUDITING

¶805 Auditing to control risks

Health and safety auditing has already been referred to in Chapter 4, where it was described as a systematic and periodic review of the whole OHS management system, including the policy and programs used to promote OHS and to prevent workplace accidents and incidents. All organisations, whether large or small, can benefit from applying safety auditing techniques. Audits may be carried out over the whole organisation, or over sections of the organisation, or in relation to a particular aspect of risk control such as electrical safety or fire safety.

Safety auditing involves evaluating safety performance using checklists and measurable criteria. It aims to determine whether the health and safety management system as a whole (the organisation's policy, objectives and targets, management program and operational controls) has been properly implemented, whether all the required elements are in place and the organisation is meeting both the objectives set in its OHS policy and legislative requirements. If the audit reveals areas of weakness, either uncontrolled risks or inoperative elements of the management system, these should be addressed.

The two essential features of a safety auditing system are that the audits be repeated at intervals and that the information collected be used.

¶806 Conducting an audit

Each safety audit should build upon previous audits. Hazards that are identified by a previous audit can be further investigated by the next. This may involve fine-tuning the checklist that relates to that particular area. The results of the initial process of hazard identification discussed in Chapter 3 may be used as a standard to plan an auditing program.

Evaluating Health and Safety Performance

If auditing over time shows a marked improvement in a particular area, it may be possible to increase the time between audits of that area.

Checklists

Checklists are the traditional method of safety auditing. Checklists allow a comprehensive approach to auditing (providing the checklist is well prepared). They also result in a standard approach to the audit, so that it can be compared with past audits.

However, checklists do have the drawback of stifling thought and discussion. Auditors should not limit themselves to the list, but should test and ask questions if they come across potential hazards.

Remember that, wherever possible, questions on checklists should be framed so that answers are given as measurable standards, rather than opinions or yes/no answers.

"What if?" method of auditing

One useful approach is to use a "what if?" method; that is, auditors ask themselves what would happen in certain hazardous situations that may occur. For example, the auditor may ask "what if this machine leaked oil on to the floor?" or "what if personnel from another department were present while a chemical shipment was being unloaded?". This method can be used to supplement checklists.

Who conducts the audit?

Safety auditing is best performed by small groups of (usually) three people. It may be useful to include an employee from the particular area being audited (possibly a safety representative). Some auditing topics (such as noise, chemical safety and ventilation) will require an expert on that topic to be part of the team.

Another option is to engage an outside consultant to conduct the audit. This may be necessary in areas that require expert knowledge. In other areas, it may be desirable to engage a consultant experienced in safety auditing to perform an initial audit, which can then be used as a base for future audits.

¶807 Scope of the audit

Safety audits can cover as many aspects of a company's operation as those responsible for the audit can think of. The following starting list provides some examples of the areas that can be dealt with and the questions that can be asked.

This list should be expanded to suit the particular organisation. Some of the more specific areas at the end of the list may not apply to all organisations. Note also the comments at ¶806 on the dangers of relying too rigidly on checklists.

Starting list for safety auditing

Health and safety policy
- Does a policy exist?
- To what extent are employees aware of it?
- Is it explained at induction?

Health and safety representatives/committees
- How much training do they receive?
- How active are they?
- How often do they meet with management?
- Which suggestions have resulted in useful changes?

Training
- Is health and safety training provided as part of induction?
- Are all employees aware of the health and safety requirements of their position?
- Which positions require particular training?

Risk Management
- Have hazards been identified, and risks assessed?
- Have risks been controlled in accordance with the hierarchy of risk control?

Medical facilities
- What staff are employed (types and numbers)?
- What tasks do they perform?
- Are employees satisfied with the services provided?
- What first aid facilities are available?

Health promotion
- What health/sporting facilities are provided?
- How many employees use these facilities?

Accident investigation
- What is the system for accident/incident investigation?
- What follow-up action has resulted from accident investigation?

Continued over

Evaluating Health and Safety Performance

Accident records and statistics
- Is there an accident recording system?
- What statistics are compiled?
- What trends do the statistics demonstrate?
- Is any use made of the information collected?
- Is the organisation's safety record improving?

Workplace inspections
- How often are inspections conducted?
- How is information from inspections used?

Housekeeping practices
- Are there instances of rubbish being left at the workplace?
- Are slip or trip hazards present?

Lighting
- Is lighting suitable for the tasks performed?

Ventilation
- Are there build-ups of atmospheric contaminants?
- How often is air-conditioning equipment inspected and cleaned?
- How effective is air movement?

Noise levels
- Has employee hearing deteriorated (audiometric testing)?
- What levels of daily noise do employees experience?

Protective equipment
- What protective equipment is used?
- Is it possible to make use of solutions further up the hierarchy of risk controls to remove the need for this protective equipment?
- How often is the protective equipment actually used?
- Is the equipment properly maintained?
- Are employees trained in the proper fitting of personal protective equipment?
- Are sunscreens, hats and shades provided for outdoor workers?

Fire safety
- What fire-fighting equipment is available?
- Is a fire-fighting team trained?
- How often are fire-drills conducted?

Emergency plans
- Is a plan in existence?
- Are the employees aware of the plan?
- Are the emergency authorities informed?

Rehabilitation program
- Has the program reduced the costs of injuries?
- Are the injured employees satisfied with the services provided?

Continued over

Chemical safety
- What chemicals or other hazardous substances are kept or generated at the workplace?
- Are MSDSs in existence and readily available?
- Do employees work in accordance with the precautions listed on the MSDS?
- Is labelling adequate?
- Are the atmospheric levels of the hazardous substances in use lower than the relevant exposure standards?

Manual handling
- Have manual handling risks been assessed and satisfactorily controlled?
- Are relevant employees trained in proper manual handling techniques?
- What instances of back injuries occurred?
- Is all suitable materials handling equipment provided?

Machine safety
- Are proper maintenance routines in place (what is the frequency of inspections)?
- Are machine guards provided?

Visual Display Units (VDUs)
- Are work stations designed in accordance with ergonomic principles?
- What instances of occupational over-use syndrome were reported?
- Are rest breaks/pause gymnastics programs in place?

Electrical safety
- Does all electrical installation comply with Australian standards?
- Is equipment inspected/tested and certified in accordance with Australian Standards?
- Are all employees working with electrical equipment adequately trained (do unsafe occurrences suggest otherwise)?

Chapter 9

An Introduction to Health and Safety Problems — Work Environments

Introduction	¶901
Working environment	
Introduction	¶902
Workplace layout	¶903
Ventilation	¶904
Temperature	¶905
Lighting	¶906
Injuries from plant	
Scope of plant/equipment injuries	¶907
Purchasing	¶908
Machine guarding	¶909
Maintenance	¶910
Danger tags and lockout systems	¶911
Hazardous substances	
Introduction — Chemical hazards/hazardous substances	¶912
Health effects	¶913
Management of hazardous substances	¶914
Control of hazardous substances	¶915
Emergency planning	¶916
Slips, trips and falls	
Scope of the problem	¶917
Causes of slips, trips and falls	¶918
Managing slip and trip hazards	¶919
Vehicle fleet safety	¶920
Fire prevention and control	
Aspects of fire prevention and control	¶921

What is fire? .... ¶922
Causes of fire .... ¶923
Hazards associated with high-piled storage .... ¶924
Hazards associated with electronic equipment .... ¶925
Warning systems .... ¶926
Fire-fighting equipment .... ¶927
Overall responsibility for fire prevention
  and control .... ¶928
Emergency evacuation plan .... ¶929
Fire-drill .... ¶930
Fire prevention and control checklist .... ¶931

¶901 Introduction

Chapters 9 to 11 take the form of an introduction to some of the more common health and safety problems that exist in Australian workplaces.

Occupational health and safety research into and discussion of each topic is a dynamic area. Since a great deal of specialist research exists on each topic, readers seeking detailed information are advised to consult specific references (limited references are provided at the end of each topic).

One useful source of information that highlights current issues is *The Journal of Occupational Health and Safety — Australia and New Zealand* (CCH Australia Limited). The Journal issues six times each year and publishes writings on subjects of ongoing interest. It draws together current research as well as encouraging wider discussion of the topics featured. Where appropriate, references to issues or articles published in the Journal are given at the end of the sections.

While not all of the problems mentioned in these chapters apply to every workplace, potential sources of their occurrence should be anticipated and either avoided or removed. Note also that specific types of workplaces may contain other problem areas not discussed in these chapters. Again, consultation of specialist references may be necessary.

The overall aims of these chapters are to acquaint readers with the main issues of each problem, and to provide an indication of what types of corrective measures are possible. The general guidelines that are provided are not a substitute for specialist assistance.

The problems described in these chapters include issues that arise from work environments, workplace illness and social behaviour. Workers compensation systems, however, distinguish between just two categories — occupational diseases and safety problems. They do this on the basis that in diseases, the harm is caused gradually (eg hearing loss or back pain), while safety problems tend to have immediate and visible consequences (eg hand injuries in machinery or slips and falls). In practice, many problems can have adverse effects on both health and safety, eg shiftwork, or drug or alcohol misuse.

During the risk management process, all risks associated with particular jobs/tasks should be considered in an integrated manner, eg risks associated with the use of a lathe in a joinery should be considered in terms of safety hazards (eg risk of hand injuries or clothing entanglement) as well as health hazards (eg noise and hazardous substances such as wood dust).

Public health issues (eg Legionnaires disease, or the impact of a lead smelter on the neighbourhood around the smelter) also need to be managed using risk management processes.

WORKING ENVIRONMENT

¶902 Introduction

The scope of occupational health and safety includes the physical working environment. This term covers aspects such as workplace layout, ergonomics, lighting, temperature, ventilation and space.

These matters tend to be of a highly technical nature. State legislation sets minimum standards for most of them. Codes of practice and Australian Standards may also apply. There may also be extra requirements for certain work environments such as bakeries, foundries, electroplating, abrasive blasting, spray painting and welding. In general, outside professional advice on these topics will be required. The purpose of the following paragraphs is to provide some indication as to whether working conditions are adequate. For sources of further assistance, refer to Chapter 10.

Work environments should be carefully considered when new office equipment is being installed and should be considered in consultation with a workplace health and safety committee and/or representative. A common example has been the introduction of visual display units. These involve a great deal of close eye work, and so require favourable lighting and ergonomic layout in particular.

Finally, although technical advice may be needed, do not hesitate to seek consultation with and suggestions from employees where a committee, or equivalent, is not established. They are the ones who have to cope with and adapt to the conditions provided, so it is appropriate that they have some say in the matter. They may also be able to alert management to potential problems and use their on-the-job knowledge to come up with valuable suggestions.

¶903 Workplace layout

It is not be possible to set out guidelines that apply to every workplace, because of the tremendous differences between work environments, dictated by type of job, equipment, and so on. However, the following summary provides some idea of which aspects to consider.

Summary — Workplace layout

1. Flexibility — so that layout can be easily changed as needs change. For example, avoid using permanent room/desk dividers when movable ones will be equally effective.
2. Layout should facilitate communication between employees, so that work flow is assisted. People whose jobs are interrelated should where possible be positioned near each other.
3. A certain amount of privacy will also be required. For example, a closed room for interviews in the Human Resources Department may be needed. Also, employees require some protection from possible sources of distraction.
4. Employees should have adequate working space, as well as furniture, storage space, etc.
5. Work stations should be designed so that employees are not repeatedly required to bend, stretch, twist their torsos or raise their arms above their shoulders. Commonly used controls, switches and tools should be placed within comfortable reach of the employee.
6. Some effort should be made to provide a pleasant office or factory landscape (for example, through use of colours, position of windows, posters, etc).
7. Some recreation space and adequate passageway space should be provided.
8. Facilities such as toilets, locker rooms, first aid room, tea-room, etc, should be adequately equipped and readily accessible.
9. Direction signs (for example, location of departments, names on doors) should be clearly and prominently installed.
10. Equipment such as air-conditioning, heating, switch room, etc, should be positioned so that their operation (noise level, etc) will not cause discomfort to any employees.
11. Safe means of entry and exit from premises is required, as well as rapid exit means in case of emergency.

¶904 Ventilation

State regulations (both safety regulations and building regulations) cover aspects of ventilation and extraction. Fresh air, supplied by either natural or mechanical means, is needed to keep down the level of carbon dioxide and to dissipate odours and contaminants.

The three main types of ventilation are natural, dilution and local exhaust ventilation.

- *Natural ventilation* is provided by outdoor air movement. Size and positioning of doors and windows is important. Natural ventilation cannot be relied upon to remove contaminants from the atmosphere.
- *Dilution ventilation* refers to the removal of old air and the replacement of it with clean or "make-up" air. It generally relies on exhaust fans removing old air from the building. Dilution ventilation is suitable where there is a low level of contamination, especially if people do not work near the source.
- *Local exhaust ventilation* works by capturing the contaminant near the point of release and removing it from the operator's breathing zone. Local exhaust systems are necessary in the case of highly toxic gases or fumes. Common types include extraction fans and exhaust hoods.

Exposure limits, which set out safe levels of contaminants in the atmosphere, are discussed at ¶913.

Ventilation for climate control

Air-conditioning is now a common feature of indoor workplaces. State regulations generally cover the construction and maintenance of air-conditioning systems. Systems should be positioned to minimise noise level problems. Maintenance must be adequate to check for problems such as Legionnaires' disease, which is caused by a bacteria that can grow in the water of air-conditioning systems.

Further references:

Managing Indoor Air Quality, Building Owners and Managers Association of Australia Limited, Sydney, 1994.

Air-conditioning and Thermal Comfort in Australian Public Service Offices, Comcare Australia, 1994.

Health and Safety in the Office, NSW WorkCover, 1993.

Officewise: a Guide to Health and Safety in the Office, HSO Victoria.

Indoor Air Quality, Division of Workplace Health and Safety, Queensland.

Recommended Indoor-Air Quality for Air Conditioned Offices in the Northern Territory, Work Health Authority, Northern Territory.

Australian Standard AS 1668, Rules for the use of mechanical ventilation and air conditioning in buildings, also associated commentary (MP47), Standards Australia.

¶905 Temperature

Discomfort at work can be caused by temperatures that are in the extreme ranges of heat or cold. Each work activity has a temperature range in which it is usually comfortable to perform the task. Temperature measurements are ideally taken at individual work stations, 1,200 mm above floor level.

State regulations may set minimum temperature limits below which radiant heating may be required to maintain temperature levels. Some occupations, however, can only rely upon warm clothing to ensure comfortable body heat (for example, cool store or refrigeration workers and outdoor workers in cold climates).

In buildings where the work carried out generates large quantities of water vapour (steam or spray), such as laundry pressing rooms and kitchens, the water vapour should be collected at the source and removed by the use of suitable enclosures, hoods and extraction systems.

In cases of high temperature, the following actions can be taken:

- introduce additional mechanical ventilation with suitable temperature control devices;
- increase air movement within the workplace;
- install evaporative coolers or air-conditioning;
- reduce entry of solar heat to building, by, for example, placing reflective material on windows, awnings, shades and blinds;
- insulate heat-generating locations with barriers and screens, and use extraction systems to remove hot air and vapours;
- reduce the periods of time for which workers are exposed to extreme conditions;
- provide cool water dispensers nearby;
- in extreme conditions, provide special protective clothing with in-built cooling system; and
- provide cool water dispensers.

Simple increase in air flow is effective in cooling the workplace where the air temperature is below 35°C. In conditions that are not too humid, evaporative coolers can be effective. In higher temperatures with humid conditions, only air-conditioning will be effective in reducing the

Work Environments

temperature, although other measures (such as increased air movement, awnings) can have some effect.

Further references:

Air-conditioning and Thermal Comfort in Australian Public Service Offices, Comcare Australia, 1994.

Thermal Comfort at Work, Department of Employment and Industrial Relations, 1981, now published through the National Occupational Health and Safety Commission.

¶906 Lighting

Lighting is another highly technical subject, also covered by State regulations, codes of practice and Australian Standards. The following guidelines will help provide employers with a useful introduction to the subject.

GUIDELINES — LIGHTING

1. Work area illumination should be suitable for all intended purposes, and reasonably uniform throughout. If a higher level is needed for particular tasks, the overall level should be increased to reduce contrast. Brightness distribution should be controlled.
2. Lighting should be installed so as not to cause glare within employees' fields of vision.
3. Colours. A white or near-white non-glossy finish on the ceiling or roof underside is preferred. Reasonably light colours are preferable for other surfaces. Remember that colours will greatly influence the lighting effect and employees will be subjected to these conditions during their full period at work.
4. Some access to natural daylight (windows, etc) is both necessary and desirable. It should be provided in conjunction with artificial lighting.
5. Alternative emergency lighting facilities should be provided in case of failure of the normal system.
6. Areas such as stairwells and escape exits should be illuminated and remain lit whenever the building is occupied.
7. Individual tasks may require special study, such as bad reflections, close visual concentration, positioning of employee, details reflected on shiny surfaces, etc. A common example is reflections on the screen of visual display units. Careful planning of office layout will be needed to overcome these difficulties.
8. Faults, such as flickering lights, blown globes, etc, should be attended to as soon as possible.
9. Sometimes a stroboscopic effect may occur when revolving machinery parts are located in proximity to light fittings. Bays of lights may remove this effect.

Further references:

Health and Safety in the Office, NSW WorkCover, 1993.

Officewise: a Guide to Health and Safety in the Office, HSO Victoria.

Artificial Light at Work (1984), *Colour at Work* (1980) and *Daylight at Work* (1983), Department of Employment and Industrial Relations, now published by The National Occupational Health and Safety Commission.

Australian Standard AS 1680.1 Interior Lighting in Buildings, Standards Australia.

INJURIES FROM PLANT

¶907 Scope of plant/equipment injuries

"Plant" is a term that includes "any machinery, equipment (including scaffolding), appliance, implement or tool and any component or fitting thereof or accessory thereto". Plant injuries include injuries involving means of transport (cars, trucks, trains), traditional industrial machinery (lathes, manufacturing equipment, printing presses), lifting equipment (cranes, fork-lifts) hand tools (power saws, staple guns) and visual display terminals. Injuries from these sources account for over 40 percent of all workplace injuries and fatalities.

¶908 Purchasing

Good engineering can provide permanent solutions to problems with plant safety. Purchasing specifications must incorporate all required safety features (such as adequate guarding of dangerous moving parts). Where sufficient expertise is not available within the organisation, it may be necessary to engage engineering and/or ergonomic consultants.

Advice from operators and maintenance staff should be sought prior to purchasing equipment. They are the people most likely to know the safety features which should be incorporated into any new machinery.

Part of any acceptance check of new equipment should be to determine that the equipment meets the safety specifications of the organisation, in addition to any legislative requirements concerning plant safety or machine guarding.

Legislative requirements affecting manufacturers and suppliers

Health and safety legislation in each State and Territory places a duty on designers, manufacturers, suppliers, erectors and installers of plant to ensure the safety of plant that they provide. This includes the adequate guarding of

Work Environments

any dangerous parts. It may also include a requirement to carry out safety testing on the plant, and to supply information relating to the plant.

Suppliers of defective plant that leads to an injury at the workplace could find themselves liable to a fine under health and safety legislation, as well as a common law action based on negligence.

¶909 Machine guarding

Where a hazard is associated with plant, it should be eliminated or isolated. When this is not possible, however, guarding may be the next preferred option. Guards need to be provided whenever there is the potential for people to come into contact with hazardous situations. So, for example, guards are required to cover very hot pipelines or to protect people's eyes during grinding processes. Guards must be provided whenever there is the possibility of a person's hands, hair, clothing, etc, getting caught in or coming into contact with moving machinery. Guards should be provided whenever there is the possibility of objects flying off moving machinery.

Some common machinery parts that require guarding include:

- flywheels connected to a moving engine;
- gearing equipment;
- belt and pulley drives;
- chain and sprocket drives;
- electric generators;
- motors;
- feed-in rollers;
- exposed electrical conductors; and
- rotating or reciprocating machine elements.

Design of guards

Where possible fixed guards should be used. These are guards that do not need to be removed for inspections and routine servicing. Where it is necessary to remove a guard for servicing, it should be secured so that it cannot be removed without the use of tools.

Guards should be sufficiently strong to support the weight of a person standing or leaning on them. They should have no sharp edges or protruding corners.

Movable guards must be interlocked with the machine, either mechanically or electrically. This means that the machine cannot be started, or cannot continue in operation, without the guard being securely in place.

The guard should not be able to be opened while the machine is in operation. Where electrical interlocking is used, it should be "fail safe" (so that the machine will shut down if something goes wrong).

Another alternative where access to the dangerous parts is required is to install a trip device (such as photoelectric beam or pressure sensitive mat). The machinery will shut down when the trip device is activated.

Legislation covering machine guarding

In each State and Territory the employer is under a general duty to provide a safe workplace for employees, including safe plant (see ¶203). Historically, there has been a legislative focus on guarding all dangerous machinery, and/or setting out guarding requirements for specific machinery. The machinery that is likely to be subject to specific requirements includes:

- chaff-cutting machines;
- circular saws and other wood-cutting machinery;
- explosive-powered tools;
- foundry equipment;
- power presses;
- steam pipes; and
- welding equipment.

There are also Australian standards dealing with the guarding of specific machinery (see "Further references" at the end of this section).

The legislation relating to guarding is tending to be replaced with new legislation based on the risk management approach (see Chapter 3) and consistent with the national standards described below.

National standards for plant and operator certification

Two national standards target health and safety related to plant and equipment use. These are the *National Standard for Plant* and the *National Occupational Health and Safety Certification for Users and Operators of Industrial Equipment*.

The *National Standard for Plant* sets out hazard identification, risk assessment, and risk control procedures for all types of plant. The certification standard specifies core competencies which are held to be the basic requirements for safe equipment use and operation. When adopted in all States, a national certification system, with certificates issued in one jurisdiction recognised by all others, will be in place.

¶909

The two standards are complementary, so that, for example, anyone undertaking work in conformance with the plant standard, such as the erection and installation of particular plant, is required to hold certification in accordance with the certification standard.

Legislation covering plant and certification of operators

Specific regulations covering plant and certification of operators apply in most jurisdictions. Some of the regulations adopt the national plant and operator certification standards declared by the National Occupational Health and Safety Commission.

As of May 2000, the common essential requirements of the national standard have been adopted by the Federal Government, Victoria, South Australia, Western Australia, Tasmania and the Northern Territory. Queensland has developed the *Plant Advisory Standard*, which commenced in February 2000, and New South Wales is expected to incorporate plant safety provisions into its forthcoming consolidated OHS regulations. The Australian Capital Territory has no fixed deadline for incorporating the common essential requirements for plant into its OHS regulations, although it has adopted the *National Standard for Plant* as a code of practice.

¶910 Maintenance

Adequate maintenance keeps plant running productively, as well as safely.

Preventive maintenance versus breakdown maintenance

There are two possible strategies regarding the maintenance of machinery:

1. Run the machinery until it breaks down, then either repair or replace it. This is known as breakdown maintenance (also on-call, unplanned or band-aid maintenance).
2. Develop a planned series of tasks, based on the likely service life of the machinery, that will keep the machine in working order. This is known as preventive maintenance.

Choosing between the two systems of maintenance is a management decision. However, this decision should not by guided purely by economic factors. The time loss associated with a machinery breakdown, as well as the possibility of injury to employees and damage to other equipment, should be considered.

Preventive maintenance

Preventive maintenance, by its nature, requires a regular pattern of tasks to be performed. It lends itself to the preparation of written procedures and/or checklists. A maintenance procedures manual should be a requirement of every major machinery purchase.

In some cases preventive maintenance will be required by legislation. Examples include boilers and pressure vessels, lifts and hoists, and air-conditioning water coolers.

Increased risks to maintenance workers

Maintenance workers will often be exposed to a higher safety risk than other workers. They will be working on or within the containment systems designed to protect other workers.

For this reason it is important to have rigidly adhered to safety procedures in place for maintenance workers. These may include a system of isolation or lock out (see ¶911). Where maintenance workers are working in areas containing dangerous chemicals, or in hot or confined areas, it may be necessary to monitor their well-being closely.

Where contract maintenance is used, it will be important that the maintenance workers know the organisation's relevant safety procedures. These could include the location of first aid facilities, fire and emergency procedures and local lock-out systems.

¶911 Danger tags and lockout systems

During the installation of new plant or maintenance of existing equipment, a lockout safety device system can be designed and incorporated into the main control mechanisms.

The lockout safety device system requires each person to be issued with their own lock and key. The keys are not interchangeable. The employee places the lock on the lockout device that is fitted to the main control mechanism and it remains there until the work is completed. The employee alone must remove the lock. Strict observance of this rule prevents accidents to those working in out-of-sight locations.

On small plant and equipment it is possible to install a lockout device or lockout main switch in the original installation, so it can be locked in the "off" position. This method is more positive than just the removal of fuses or an iron bar through the hand wheel of a valve.

A danger tag system does not offer as positive a means as the previously mentioned isolating devices, but it is designed to offer a measure of personal protection on existing plant and equipment without major capital outlay for modifications. Also, the danger tag system can be effectively introduced to promote greater safety consciousness amongst the workforce and does offer a start in providing a definite hazard spotting system within the working procedures.

An example of tags is shown below.

The danger tag system should embrace all situations in which danger to persons could arise from the operation of machinery, plant or equipment, by the flow of steam, electricity, gases or liquids, or the use of faulty or unsafe plant and equipment. As a bonus, the system can minimise further damage to plant and equipment caused by inadvertent starting.

Danger tags have a wide range of applications, including large machinery and equipment in factories, "field" sites, and small items of plant and equipment, such as tools, ladders or defective vehicles (in the latter case, the tag could be attached to the steering wheel).

The system could embrace a control return point where defective portable items can be returned for repair to a workshop, thus speeding up the procedural flow and minimising the time that the unit is out of service. This allows plant and equipment to be used to its fullest degree, provides for maximum efficiency and minimum duplication of units.

Further references:

Plant in the Workplace: Making it Safe, National Occupational Health and Safety Commission, 1995.

Plant Design: Making it Safe, National Occupational Health and Safety Commission, 1995.

Core Training Elements for the National Standard for Plant, National Occupational Health and Safety Commission, 1995.

National Occupational Health and Safety Certification for Users and Operators of Industrial Equipment, National Occupational Health and Safety Commission, 1995.

National Standard for Plant, National Occupational Health and Safety Commission, 1995.

Controlling Risks from Plant (Machinery and Equipment) , ACTU, 1996.

Standards Australia publishes a range of standards on the guarding of specific machinery, including: *AS 1893 (guillotines); AS 1473 (woodworking machinery); AS 1788 (abrasive wheels); AS 2153 (agricultural tractors and machinery); AS 1873 (explosive-powered fastening tools); AS 1219 (metal power presses); and AS 1895 (portable electrical tools); AS 4024 (safeguarding of machinery)*.

HAZARDOUS SUBSTANCES

¶912 Introduction — Chemical hazards/hazardous substances

Thousands of different chemicals are used in industry, with new ones being introduced at a global rate of 1,000 to 2,000 per year. The properties and health effects of many of these chemicals are well known, but it is not possible to obtain comprehensive information regarding all industrial chemicals. The asbestos disaster reminds us that the health effects of some substances may not become apparent until after many years of exposure.

However, it is possible to minimise the risks associated with chemicals and other hazardous substances at the workplace by adopting certain procedures. These procedures revolve around two aspects:

1. adequate information, readily available; and
2. control measures to eliminate or minimise the risk to workers.

These two aspects are developed in the following paragraphs.

The main hazards from chemicals are:

- environmental pollution;
- fire/explosion, or other physical damage; and
- health effects.

This commentary will concentrate on the health effects of hazardous substances at the workplace, with some discussion of precautions against emergencies such as fires (see ¶916). However, an organisation planning its hazardous substances management strategy will have to take all three forms of hazard into account.

Work Environments

The definition of a hazardous substance is a substance that meets either or both of the following criteria:

- It is listed in the National Occupational Health and Safety Commission's *Designated List of Hazardous Substances*.
- It meets the requirements of the National Commission's *Approved Criteria for Classifying Hazardous Substances*.

In general, the term "hazardous substances" is used to indicate the potential to harm health — with the harm being done over time. In this way, "hazardous substances" are distinguished from "dangerous goods". The latter are generally assumed to pose an immediate risk to personnel, property or environmental safety, in terms of explosion, fire or poisoning. There is significant overlap between these two categories, such that many hazardous substances are also classified as dangerous goods. Separate legal requirements apply to dangerous goods; and where an overlap occurs, both the hazardous substances requirements and the dangerous goods provisions apply.

National chemical assessment scheme

The National Industrial Chemicals Notification and Assessment Scheme (NICNAS) came into effect in July 1990. Under the scheme, all new and some existing chemicals imported into or manufactured in Australia are required to be notified and assessed. Each chemical is assessed for its potential effects on occupational health, public health and the environment. The scheme is administered by the National Occupational Health and Safety Commission.

¶913 Health effects

Different chemicals cause a wide range of health effects.

> *Acute poisons* are those where a short exposure causes an immediate health effect. Some chemicals, such as strong acids, have an immediate effect on contact with the skin or eyes, while others, such as cyanide, have a rapid effect on inhalation.
>
> *Chronic toxins* are chemicals that cause harm due to prolonged exposure. For example, build-up of mercury in the body can cause nerve and brain poisoning. Note that many chronic toxins will become acute poisons in large doses.
>
> *Corrosive chemicals and irritants* usually affect mucous membranes (mouth, nose, throat, lungs). Inhalation of a high dose of sulphuric acid fumes will cause lungs to collapse and possibly death.

Allergens or sensitisers cause a change in the body's reaction to that substance, so that subsequent exposure causes a reaction, most notably dermatitis or asthma. For example, isocyanate compounds can cause sensitisation resulting in asthma-like symptoms. It is not usually possible to tell who will become sensitised to a particular chemical. Occupational skin problems, often caused by allergens or sensitisers, are discussed at ¶1010 et seq.

Carcinogens are those chemicals that are capable of causing cancer. They include asbestos, synthetic mineral fibres, arsenic compounds, benzene and vinyl chloride. Occupational cancer is discussed in more detail at ¶1005 et seq.

Mutagens change the genetic code of cells and can cause mutations.

Teratogens affect the growth of the fertilised egg and embryo, also causing mutations.

Routes of entry

In industry, there are three main ways in which chemicals can enter the body.

1. Inhalation: breathing chemicals into the lungs; the most common method of chemicals entering the body at work.

2. Body contact: chemicals coming into contact with the skin or eyes should generally be avoided.

3. Ingestion: eating or drinking; usually not such a problem, but hands should be washed before eating or smoking to avoid accidental ingestion.

Monitoring doses

The risk of adverse health effects depends on the dose of a chemical that is absorbed into the body. For this reason it is important to monitor the concentration of dangerous substances in workroom air. These measurements require a knowledge of occupational hygiene as well as special equipment. The tests must be done in areas where workers are located, and they must be done so that both typical and worst levels are recorded.

These measurements can then be compared with *Exposure Standards* determined by various health and safety bodies. *Exposure Standards* establish a level of exposure to which, it is believed, virtually all workers may repeatedly be exposed without adverse effect. The National Occupational

Health and Safety Commission has issued such standards for more than 650 chemicals.

One commonly used *Exposure Standard* is known as a *Threshold Limit Value* (TLV). Often a TLV is issued as a *Time Weighted Average* (TLV-TWA), which refers to the safe limit of exposure averaged over eight hours. This can be exceeded for a short period, provided the exposure does not exceed the Short-Term Exposure Limit (TLV-STEL), which usually refers to the safe limit for 15 minutes of continuous exposure. More dangerous chemicals will not have a Time Weighted Average, but a Ceiling TLV, which should not be exceeded at any time.

¶914 Management of hazardous substances

The first step in any hazardous substances management program is to establish an inventory of all chemicals used at the workplace. Where this is not already in place, it will involve a survey of the chemicals and other hazardous substances in the workplace, as well as establishing a procedure for recording new substances as they enter or are produced by the work process at the workplace. The questions to be asked include:

- What materials are present?
- Where are they located?
- How are they stored?
- How much is stored at each location?
- How are they used?
- How often are they used?

All areas of the workplace must be considered, including laboratories, receiving areas, offices and maintenance sections.

Material Safety Data Sheets

Once it is known which chemicals are present at the workplace, it is necessary to obtain information on those chemicals. This can be done by obtaining a Material Safety Data Sheet (MSDS) for each chemical. These data sheets detail relevant health and safety information on a substance. It should be possible to obtain an MSDS from the supplier or manufacturer of the substances. If the business is producing or generating a hazardous substance, it may be necessary to compile an MSDS for it.

The National Occupational Health and Safety Commission (NOHSC) has published a National Code of Practice for the preparation of Material Safety Data Sheets, which should be used when compiling MSDSs. Legislation requires that suppliers and manufacturers of substances must supply this information. It is worthwhile insisting that all materials coming into a workplace are supplied with an MSDS, or at least with a list of ingredients and safety information. However, it will then be necessary to "customise" the MSDS, by adding information such as the use to which the substance is put, and the place(s) where it is stored. Safety hints discovered when using the substance should also be added to the MSDS, so that they are not lost when new personnel come into the workplace.

Additional information for completion of MSDSs can be obtained from health and safety bodies. NOHSC has established a national health and safety database, containing sample MSDSs, approved exposure standards and information on the toxic effects of many chemicals.

The list of chemicals stored at the workplace and the attached MSDSs should be stored in a filing system with easy access. Inexpensive software is available to perform this task.

The sample MSDS opposite is reproduced with permission of the Chamber of Minerals and Energy of Western Australia, Inc.

Labelling

Another important aspect of hazardous substances management is adequate labelling. It is not possible to provide all safety information on a label, and it is better if the label contains minimal information but enables the user to refer easily to the correct MSDS. A clear substance name is the major function of a label, with perhaps an identifying number and mention of any particularly dangerous ingredients. Labelling should conform with the NOHSC's Code of Practice for the Labelling of Workplace Substances.

MATERIAL SAFETY DATA SHEET

Product	Noricene 8940

Manufacturer Details: Nalco Australia Pty Ltd
[Address, telephone, Fax]

SYNONYMS: None

UN: 1789   **CLASS 8**   **SUB RISK:**   **HAZ CHEM CODE:** 2R
EPG: zz38   **PACKAGING GROUP:** II   **POISON SCHEDULE:** 6

USES: Inhibited hydrochloric acid descaler.

PHYSICAL AND CHEMICAL PROPERTIES

APPEARANCE: Clear amber liquid with a pungent odour.

BOILING POINT: 84°C   **MELTING POINT:** 0°C
VAPOUR PRESSURE: 35 mmHg   **VAPOUR DENSITY:** 1.15
SPECIFIC GRAVITY: 1.27   **SOLUBILITY (WATER):** Soluble
FLASH POINT: None   **EXPLOSION LIMITS:** None
% VOLATILES: Negligible   **pH:** 1% Solution: 1

INGREDIENTS

Chemical Name	CAS Number	Proportion
Hydrochloric acid	7647-01-0	44%
Proprietary corrosion inhibitor		<10%

HEALTH HAZARD INFORMATION

INGESTION: Corrosive, may cause burning of the mouth, throat and stomach. Ingestion may cause coughing, constriction of the throat due to swelling of the larynx and vomiting of blood or blood in diarrhoea. Breathing difficulties, shock and convulsions may follow.

EYE: Vapour is severely irritating. Liquid is severely irritating and may cause conjunctivitis, ulceration and corneal burns. Permanent eye damage may result.

SKIN: Irritating after brief contact. Prolonged contact may result in inflammation and burns.

INHALATION: Product has relatively low volatility so inhalation of hazardous quantities of vapour is unlikely to occur during normal use. However, if generated, inhalation of vapours or spray mists are severely irritating. Inhalation may cause burning sensations, higher concentrations may cause coughing, choking and difficulty in breathing due to oedema of the lungs. Symptoms of oedema may have delayed onset.

CHRONIC: Hydrochloric acid
    LCLo (inhaled, human): 1300 ppm/30M
    LD50 (oral, rabbit): 900 mg/kg
    LCLo (inhaled, rabbit): 4413 ppm/30M

Inhalation and ingestion are the routes of entry into the body. Adverse health effects are related to corrosive nature. The estimated fatal dose for an adult is 1 mL of concentrated acid.

FIRST AID PROCEDURES

INGESTION: NEVER GIVE AN UNCONSCIOUS PERSON ANYTHING TO DRINK NOR ATTEMPT TO INDUCE VOMITING. If the person is conscious, rinse mouth out with water ensuring that mouth wash is not swallowed. Give about 250 mL (2 glasses) of water to drink. DO NOT attempt to induce vomiting. Seek URGENT medical attention.

EYE: IMMEDIATELY hold eyelids open and rinse the eye continuously with a gentle stream of clean running water for at least fifteen minutes. Seek URGENT medical attention. If practical, continue irrigation of the eye during transportation to medical facilities.

SKIN: Remove contaminated clothing. Rinse the affected area with water then wash thoroughly with soap and water. Use water alone, if soap is unavailable. Seek medical attention if any soreness or inflammation of the skin persists. Launder affected clothing before re-use.

continued over ...

¶914

INHALATION: Remove to fresh air. Keep warm and at rest. If breathing is laboured, hold in a half upright position (this assists respiration). Apply artificial respiration if breathing has stopped. Seek URGENT medical attention for all but the most minor cases of over-exposure.

PRECAUTIONS FOR USE

ENGINEERING CONTROL: Ventilation requirements depend on the concentration of solution in use, the quantity and the method of application. General (mechanical) ventilation is adequate for minor use. Local exhaust ventilation may be required if handling large quantities of concentrated product.

FLAMMABILITY: Non-combustible. May evolve acrid fumes if heated. Concentrated solutions react vigorously with alkalis. May evolve hydrogen on contact with metals.

PERSONAL PROTECTION: Requirements are dependent on working conditions, quantity of product in use and its concentration. For minor use, acid resistant safety goggles and rubber gloves may be adequate. If concentrated solutions or large quantities are used, [wear] acid resistant safety goggles, face shield, gloves or gauntlets, overalls and splash apron. A half face respirator with acid gas filter is required if the product is being sprayed or handled in such a way that hydrogen chloride vapour is generated. In confined spaces use air supplied breathing apparatus. N.B. TAKE THE LIMITS OF ABSORPTION CAPACITY INTO ACCOUNT. CHANGE FILTERS REGULARLY.

EXPOSURE STANDARD

E.S. CHEMICAL NAME	TWA ppm mg/m³	STEL ppm mg/m³
Hydrochloric acid	5.00 7.50	Peak

T.L.V. CHEMICAL NAME	TWA ppm mg/m³	STEL ppm mg/m³
Hydrochloric acid		5.00 7.50

SAFE HANDLING PROCEDURES

STORAGE TRANSPORT: Store in a cool well ventilated area. Avoid contact with reactive metals. Keep separate from alkaline substances. Protect containers against physical damage.
Class 8 Corrosive Substances should not be transported or stored with goods of:

Class 1	Explosives
Class 4.3	Dangerous When Wet Substances
Class 5.1	Oxidising Agents
Class 5.2	Organic Peroxides
Class 6	Poisonous (toxic) Substances (where the poisonous substances are cyanides and the corrosives are acids)
Class 7	Radioactive Substances

Foodstuff and foodstuff empties

SPILLS DISPOSAL:
SPILLS: Dilute solutions and very minor spills may be flushed away with water. For major spills or those involving concentrated product wear full protective equipment including air supplied breathing apparatus, gloves, acid resistant overalls and rubber boots (trousers should be worn OVER boots, not tucked in). If possible the spill should be contained by damming with earth or sand and then covered with a weak alkali such as soda ash or sodium bicarbonate. It can then be flushed away with copious quantities of water.
DISPOSAL: Dilute solutions: flush to drains with plenty of water. Concentrated solutions: dilution and chemical reaction or to approved land-fill.

FIRE EXPLOSION: Non-combustible but may evolve acrid fumes in a fire situation. May evolve hydrogen, a highly flammable gas, on contact with reactive metals.
Wear self contained breathing apparatus. Extinguish using whatever is suitable for the primary cause of the fire. Water sprays may be used to 'knock down' vapour.

USER COMMENTS

¶914

Work Environments

¶915 Control of hazardous substances

There are various methods of controlling chemical hazards at the workplace. These can be split into two basic groups:

1. engineering and design methods; and
2. controls which depend on modification of worker behaviour.

Engineering and design methods are more effective in reducing chemical hazards and should be considered before behavioural controls.

Engineering and design controls

1. *Elimination or substitution.* Where possible, hazardous substances should be eliminated from the particular process or substituted with a less dangerous chemical. It may also be possible to substitute a less dangerous process involving the same chemical, such as using dipping rather than spraying. A good technical knowledge of the particular industry/process is required to determine whether elimination or substitution is possible.

2. *Isolation or enclosure.* It may be possible to isolate the majority of workers from a dangerous process, or alternatively to enclose the process completely. For example, many solvent degreasing units are designed so that the degreasing solvent is completely enclosed.

3. *Ventilation.* This can be used to control airborne contaminants. Local exhaust ventilation, where the contaminant is sucked away from the point of release, is the most effective. Dilution ventilation relies on clean air being pumped into the workplace and the old air being released. Natural ventilation is the least effective method.

Controls that rely on human behaviour

1. *Information and training.* Workers should understand the hazards that they face, be able to access and interpret MSDS and know the procedures they should follow to minimise the risks.

2. *Supervision.* Good supervision is necessary to ensure that work is done safely, and problems are addressed as they arise. Supervisors need to be trained to recognise unsafe systems of work.

3. *Monitoring.* Monitoring is necessary to ensure that particular risks are under control. Monitoring is discussed above in the context of Exposure Standards (see ¶913).

4. *Personal protective equipment.* This is the least preferred option. Personal protective equipment (gloves, respirator, protective

clothing) should be used only where it is not possible to introduce engineering or design controls. Protective equipment is discussed in more detail at ¶629.

Legislative provisions, codes and standards

In 1994, NOHSC declared a national regulatory reform package for the control of workplace hazardous substances. The package included *National Model Regulations for the Control of Workplace Hazardous Substances*, as well as codes of practice for the preparation of material safety data sheets, for the labelling of workplace substances, and for the control of workplace hazardous substances. Part 2 of the National Commission's code for the control of workplace hazardous substances relates to scheduled carcinogenic substances and was published in October 1995.

Since then, all States and Territories have adopted hazardous substances regulations consistent with the National Commission's model regulations and codes of practice (the *Victorian Occupational Health and Safety (Hazardous Substances) Regulations 1999* commenced on 1 June 2000).

¶916 Emergency planning

A final consideration in relation to chemical safety is planning for emergencies. This relates particularly to substances regarded as dangerous goods. It involves planning for the safest storage of chemicals, and placarding dangerous goods in order to be of assistance to emergency services.

When planning a new installation, it is worthwhile to consider the sorts of emergencies that may arise. Flammable chemicals, particularly, should be stored separately from the main workplace.

Placarding

All chemical storage areas should be placarded in order to provide information to emergency services. This means placing clearly visible placards setting out the name of the substance, any standard safety signs (flammable liquid, flammable solid, poison gas, oxidising agent, corrosive, etc) and the Hazchem Code for that substance. Hazchem Codes are a system, developed originally by the UK Fire Brigade, for telling the fire services certain information such as the type of fire extinguisher that should be used, what sort of personal protection is necessary and whether evacuation should be considered.

The National Occupational Health and Safety Commission has published a *Guidance Note on the Storage of Chemicals*, which details the sorts of placards that are necessary. Specific placards should be placed on each

Work Environments

chemical in bulk storage, while composite placards (setting out the type of chemicals contained within) can be placed at the entrance to a mixed storage. Examples of placards are shown below.

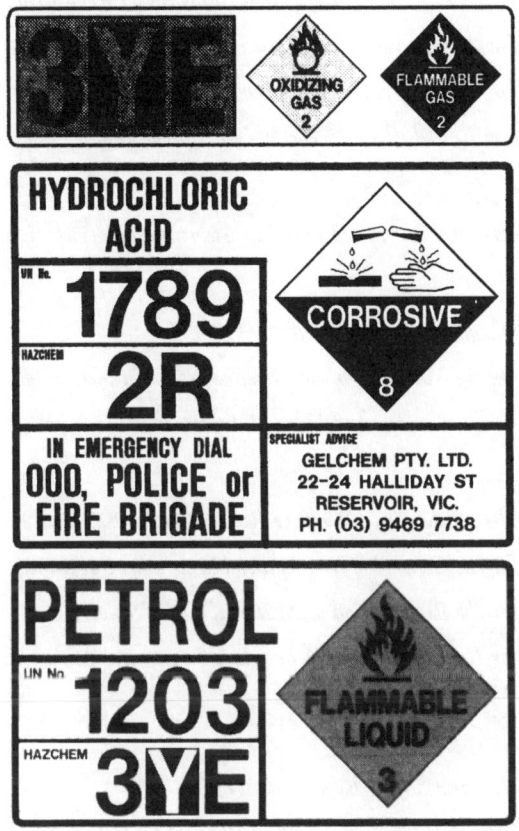

Placarding is also required on dangerous goods that are being transported. These requirements are set out in the *Australian Code for the Transport of Dangerous Goods by Road and Rail* (ADG Code). The ADG Code has been adopted by all States and Territories. The code applies to a large number of dangerous substances, which it splits into various classes depending on their properties.

Chemical plan/manifest

A plan of all hazardous chemicals stored at the workplace (sometimes referred to as a manifest) also assists emergency services. The plan should be kept in a readily accessible location, and should set out all the chemicals on site by class, quantity and location.

¶916

State/Territory legislation and codes of practice

NSW:

Occupational Health and Safety (Hazardous Substances) Regulation 1996
Code of Practice for the Control of Workplace Hazardous Substances
Code of Practice for the Preparation of Material Safety Data Sheets
Code of Practice for the Labelling of Workplace Substances

Vic:

Occupational Health and Safety (Hazardous Substances) Regulations 1999

Qld:

Workplace Health and Safety Regulation 1997
Code of Practice for the Management of Hazardous Substances
Code of Practice for the Storage and Use of Chemicals at Rural Workplaces

SA:

Occupational Health, Safety and Welfare Regulations 1995
Code of Practice for the Control of Workplace Hazardous Substances
Code of Practice for the Preparation of Material Data Sheets
Code of Practice for the Labelling of Workplace Substances AS 2030 (known as the SAA Gas Cylinders Code)

WA:

Occupational Health and Safety Regulations 1996
National Code of Practice for the Control of Workplace Hazardous Substances
National Code of Practice for the Preparation of Material Data Sheets
National Code of Practice for the Labelling of Workplace Substances

Tas:

Workplace Health and Safety Regulations 1998

ACT:

Approved National Model Regulation as a Code of Practice
National Code of Practice for the Control of Workplace Hazardous Substances
National Code of Practice for the Preparation of Material Data Sheets
National Code of Practice for the Labelling of Workplace Substances

¶916

NT:

Work Health (Occupational Health and Safety) Regulations 1992

National Code of Practice for the Control of Workplace Substances

National Code of Practice for the Labelling of Workplace Substances

National Code of Practice for the Preparation of Material Data Sheets

AS 2507 The Storage and Handling of Pesticides

Further references:

Storage of Chemicals, Guidance Note, National Occupational Health and Safety Commission, June 1990.

Exposure Standards and Guidance Note for Atmospheric Contaminants in the Occupational Environment, National Occupational Health and Safety Commission, May 1995.

In addition to the above, the National Occupational Health and Safety Commission publishes a number of codes of practice and guides dealing specifically with the more common hazardous chemicals, including inorganic mercury, benzene, glutaraldehyde, ethylene dioxide, organic and inorganic lead, asbestos, vinyl chloride, timber preservatives, arsenic, cyanide, hydrogen fluoride, industrial organic solvents, foundry chemicals, isocyanates, solvent vapour degreasers and welding fumes.

Controlling Hazardous Substances, ACTU Occupational Health and Safety Unit, 1995.

Control of Workplace Hazardous Substances: National Model Regulations and National Code of Practice, National Occupational Health and Safety Commission, 1994.

National Code of Practice for the Preparation of Material Safety Data Sheets, National Occupational Health and Safety Commission, 1994.

National Code of Practice for the Labelling of Workplace Substances, National Occupational Health and Safety Commission, 1994.

Approved Criteria for Classifying Hazardous Substances, National Occupational Health and Safety Commission, 1994.

List of Designated Hazardous Substances, National Occupational Health and Safety Commission, 1994.

National Model Regulations for the Control of Scheduled Carcinogenic Substances, National Occupational Health and Safety Commission, 1995.

National Code of Practice for the Control of Scheduled Carcinogenic Substances, National Occupational Health and Safety Commission, 1995.

¶916

Core Training Elements for the National Standard for the Control of Workplace Hazardous Substances, National Occupational Health and Safety Commission, 1995.

Guidance Note for the Assessment of Health Risks Arising from the Use of Hazardous Substances in the Workplace, National Occupational Health and Safety Commission, 1994.

Guidance Note for the Control of Hazardous Substances in the Retail Sector, National Occupational Health and Safety Commission, 1994.

Exposure Standards for Atmospheric Contaminants in the Occupational Environment, National Occupational Health and Safety Commission, 1995.

Guidelines for Laboratory Personnel Working with Carcinogens or Highly Toxic Chemicals, National Health and Medical Research Council, 1990.

A Guide to the Hazardous Substances Regulation and Control Code of Practice, WorkCover NSW, 1996.

Understanding Hazardous Substances in the Workplace, WorkCover NSW, 1996.

Managing Chemical Hazards in the Workplace, WorkCover NSW, 1996.

Australian Hazardous Substances Legislation, CCH Australia, Sydney, 1996.

Hazard Alert: Managing Workplace Hazardous Substances, CCH Australia Limited, 1995.

Laboratory Safety Manual, CCH Australia Limited, 1990.

Identifying Hazards at Work, ACTU Occupational Health and Safety Unit.

Chemicals at Work, ACTU Occupational Health and Safety Unit.

Standards Australia has produced many standards relevant to a particular hazard or situation.

HB76 Dangerous Goods — Initial Emergency Response Guide, Standards Australia, Sydney.

Emergency Response Manual, Australian Chemical Industry Council, 1987.

Chemical Incidents Emergency Telephone Contacts, State Pollution Control Commission.

Industrial Emergency Planning Guidelines, NSW Department of Planning.

SLIPS, TRIPS AND FALLS

¶917 Scope of the problem

In the workplace context, "slips, trips and falls" is a general term referring to accidents involving:

- falls on the same level; and
- stepping on an object.

It does not include falls from a height.

The falls resulting from slips and trips result in a wide range of injuries including back strains as well as injuries to other parts of the body. Incidents of this type in Australian workplaces are major causes of permanent disability. Figures published by WorkCover NSW for the year 1994-95 revealed that almost four in every 1,000 males employed and more than two in every 1,000 female workers suffered serious slip and trip injuries. These injuries accounted for 14.6% of the total number of serious injuries to working males and 20.3% of the serious injuries to working females. Falls on the same level are responsible for nearly 11% of back injuries, and those injuries are likely to be more severe than those resulting from over-exertion.

The average cost of compensating a slip, trip and fall accident is about $12,000, and businesses can incur additional costs in recruiting and training a new worker while the injured worker recovers, and in lowered productivity.

If the reported instances of non-employees injured in workplaces were to be added to the employee statistics (one large retailer, for example, reports that about 80% of its customer injury claims are associated with slip and fall accidents), the losses resulting from slips and trips injuries would be even larger.

¶918 Causes of slips, trips and falls

Slips occur when a person loses their footing or balance when their feet unintentionally slide over the surface they are in contact with. Slips may also include instances where other bodily parts unintentionally slide on some surface that is bearing body weight (such as the fingers losing their grip on the rung of a ladder).

Trips occur when a person loses their footing or balance when they unintentionally stumble (often after catching their foot on some object that momentarily prevents their foot from moving in the direction in which the body weight is being transferred.)

Slips and trips can occur in many different ways. Slips in the workplace result from circumstances such as:

- surfaces having poor friction characteristics (particularly when they are wet);
- surfaces being dirty, or having some slippery substances or material on them, such as lubricating oil, foodstuffs and beverages, product, fine dusts, water and rollable solids (such as beads, pellets, spherical ball bearings, pipes and regular-shaped objects);
- people wearing unsuitable footwear such as shoes with soles that have no patterned grip, are worn smooth, or have poor frictional characteristics;
- surfaces being inclined or irregular;
- people walking on surfaces where there are no handrails or balustrades; and
- people running on polished or smooth (low friction) surfaces.

Trips in the workplace result from circumstances such as:

- floor surfaces being rough or uneven;
- floor coverings, particularly carpet or matting, being torn or uneven;
- hazards where people may catch the toe or heel of their footwear in;
- some obstruction (such as trailing electrical cables or objects left on the floor);
- stairs having uneven "riser" or "going" dimensions, or stairs that fall outside the parameters of safe design (as established in *AS 1657-1992*);
- variations in floor levels that are difficult for people to appreciate or perceive (often because of poor lighting); and
- people wearing inappropriate footwear (such as high heels on grid-mesh flooring).

¶919 Managing slip and trip hazards

Most Australian OHS legislation has requirements for the design, installation and maintenance of floors, passageways and stairs that existed prior to risk management based provisions. These older provisions require the owner or occupier to meet prescribed standards. Several OHS statutes have now called up standards such as *AS 1657:1992*, the *SAA Code for fixed*

Work Environments

platforms, walkways, stairways and ladders. Employers and occupiers bear the main obligations, although the designer also has certain responsibilities.

Aside from any specific or prescriptive requirements, all employers are under a general duty to provide a safe place of work, and this broad obligation may be taken to include the duty to prevent injury resulting from slip or trip hazards.

The primary means of achieving this is to undertake risk management in relation to all slip and trip hazards in the workplace — that is, to:

- identify all slip and trip hazards;
- assess the associated risks; and
- control those risks.

There is also a need to ensure that sufficient monitoring and review is undertaken of risk management systems to ensure they remain effective, and to keep records of the risk management process (see Chapter 3). Further details on managing slip and trip hazards are provided below.

Principles for controlling risks from slips and trips

There are many ways in which the risks associated with slip and trip hazards can be controlled. These include:

- the use of high grip floor coverings;
- provision of collection points for liquid spills; and
- the enclosure of machine parts that generate dust, oil or other waste substances, so that any spillage or leakage is retained.

The elimination of slip and trip hazards must be a primary aim of workplace and plant designers and those in control of workplaces. The risk of injury associated with slip and trip hazards can be effectively minimised by:

- selecting/specifying floor surface and paving materials that give good frictional grip;
- ensuring that any wet areas where liquid may accumulate will drain to a collection point. A commonly encountered problem with wet floor areas is that they are laid flat leading to liquid "ponding" on the floor rather than flowing away from trafficable areas;
- enclosing equipment that generates dust, oil or any other substances that may foul trafficable surfaces;
- locating sources of dust, oil and other substances that may present slip or trip hazards in a bounded area;

¶919

- designing all stairs and ladders so that they fully conform with the requirements of *AS 1657-1992*;
- avoiding variation in floor levels;
- providing clear accessways that are free of obstructions and objects protruding into walking zones;
- clearly delineating the designed accessways;
- providing adequate lighting (including emergency lighting) over areas where people are required to walk;
- providing suitable handrails and balustrades; and
- using techniques such as "tiger taping" to make any unavoidable differences in floor level as visible and obvious as possible.

Manufacturers' duties

The manufacturer of plant items can play a role in ensuring that slip and trip hazards are not built in as part of the manufacturing or assembly process. Flooring, in particular, should be fitted and finished well to avoid any inconsistency or unevenness. Cabling and piping should be secured so that it cannot accidentally snag feet in tight spaces (such as in the operator's cab of mobile plant).

Installers' and erectors' duties

The installer/erector of plant must ensure that during the construction, erection, installation and commissioning stages, slip and trip hazards are identified and either eliminated or, at least temporarily, isolated (for example, by the erection of barricades or fencing). Where no specification for floor finishes in pedestrian areas is given, the opportunity may exist to introduce additives to floor finishes (such as carborundum) to increase the frictional grip of the flooring.

Owners' and occupiers' duties

The owner or occupier of premises where plant is installed must identify slip and trip hazards as part of the risk management process for the plant. The ongoing control measures adopted should include regular inspections of the workplace by local management to identify any slip or trip hazard that may have been overlooked in the initial risk management process, or that has been introduced since the last inspection.

Further information:

Preventing slips, trips and falls — guidance note, WorkCover NSW, November 1998.

¶919

Avoiding slips, trips and falls — information for workers, WorkCover NSW, November 1998.

¶920 Vehicle fleet safety

The definition of on-the-job safety must include road safety when employees are using company vehicles to carry out their work or travel to or from work.

In practice this may be a difficult aspect of work safety to control, for the following reasons:

- it is hard to supervise and maintain contact with employees when they are out on the road;
- many aspects of conditions on the road are outside employers' control; and
- control of the fleet will involve several people, such as a purchasing manager, fleet manager and maintenance staff, meaning that a health and safety officer may have to operate through several channels.

Nevertheless, there are several ways in which management can improve vehicle fleet safety. These are summarised in the table below.

Summary — vehicle fleet safety

1. *Policy* — Management could emphasise road safety in its health and safety policy, or in a separate program, setting out areas of responsibility and accountability.
2. *Driver selection* — The recruitment process could assess factors such as experience on other jobs, attitude towards safety, past driving record, references, licence check, medical examination, knowledge of vehicles, maintenance and road rules, and a road test.
3. *Driver training* — The induction process could emphasise road safety (including issue of program and traffic rules). The importance of customer relations as well as personal and vicarious liability could also be mentioned. Instruction courses could be given in defensive driving, emergency procedures, reporting accidents/breakdowns, and how to check vehicles for safety.
4. *Maintenance and record-keeping* — The maintenance arrangement may depend on the agreement with the vehicles' lessors. Records must clearly and efficiently show when maintenance is needed, what maintenance is required, record of completion, costs, vehicle condition reports and history cards, accident records (causes and circumstances) and driver records (tests, training, appraisals, complaints, accidents). Spot checks of vehicles could also be made periodically. Any breaches of set standards could be penalised in some way, such as by loss of the private use of the vehicle for a set period.

FIRE PREVENTION AND CONTROL

¶921 Aspects of fire prevention and control

The impact of a fire on an organisation can be enormous. Lives can be lost, property and stock damaged or ruined, sales lost and business disrupted while rebuilding takes place. In some cases, the business may even be unable to recommence operations. For these reasons it is in all employees' interests to observe fire prevention and control measures.

Proper provision for fire prevention and control must include each of the following aspects, all of which are outlined later in this section:

- observance of legislation (Acts, Regulations and Ordinances) relevant to fire prevention and control;
- attention to workplace "housekeeping" to reduce the chances of fire occurring;
- observance of advice from the State fire department and insurance fire surveyors;
- special attention to high density storage areas and electronic equipment;
- provision of suitable fire controls or warning devices, such as smoke detectors, heat detectors and alarm systems;
- provision of fire-fighting devices, including sprinkler systems, hydrants and hose reels;
- an evacuation plan;
- fire-drills at regular intervals;
- training of personnel in the use of fire-fighting equipment and operation of an evacuation plan; and
- sources of information on technical aspects which affect fire safety at the workplace, such as the controlled use of flammable or combustible materials.

State Regulations and Ordinances set out many provisions on fire prevention and control, from the initial building design to requirements concerning explosives and dangerous goods. State Regulations generally cover the provision of extinguishers and warning devices. Other Regulations cover specific work processes, such as spray painting and the dipping of articles in flammable solutions.

¶922 What is fire?

Basically, fire is the result of the three factors required to support combustion, which are shown by the diagram below:

1. FUEL — solid, liquid or gas;
2. OXYGEN — that exists in the air; and
3. HEAT — the ignition temperature will vary according to the material.

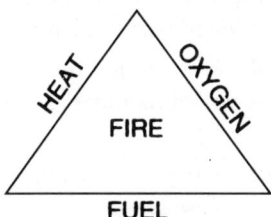

By removing one of these factors combustion cannot be supported (that is, by removing one side, the triangle will collapse).

By considering the triangle it can be seen that there are three methods of extinguishing a fire:

1. STARVATION — removal of the fuel or of combustible material in the vicinity of the fire to isolate it;
2. SMOTHERING — reducing the oxygen content of the atmosphere in the immediate vicinity of the fire by excluding air; or
3. COOLING — lowering the temperature of the burning material below its ignition temperature.

The stages of fire

The development of a fire passes through four separate stages:

1. INCIPIENT — no visible smoke or flame or significant heat, but there is a condition generating combustion particles through chemical decomposition not visible to the human eye and develops over extended period;
2. SMOULDERING — particles start to become visible as "smoke";

3. FLAMES — ignition point has occurred and infra-red energy is given off by flames visible smoke level decreases and more heat is developed; and

4. HEAT — this last stage develops very quickly from stage 3, with large amounts of heat, flame, smoke and toxic gases produced.

¶923 Causes of fire

There are a number of potential causes of fires at a workplace:

1. management failure to incorporate fire prevention and fire control as an integrated part of planning and procedures;
2. actions by people, such as failure to extinguish a cigarette, or smoking in areas where flammable or combustible materials are present;
3. conditions, particularly overloading electrical circuits and wiring, or incorrect fuse sizes;
4. electrical malfunctions, such as in motors, switchboards or generators;
5. unsuitable storage, such as combustible material near exhaust or steam pipes or high temperature areas;
6. friction, such as that produced by conveyor belts or bearings;
7. chemicals and liquids with low flash points that evaporate at room temperature to create an explosive vapour/air mixture, flammable gases and vapours, cleaning solvents and various dusts (flour/wood dust); and
8. unsuitable location of work processes and equipment.

Each of these aspects should come under scrutiny when the health and safety survey (see ¶401) is being carried out. Many causes on the above list may be regarded as "housekeeping" matters and can be included as part of the accident prevention program. All are preventable and can best be overcome by regular maintenance and inspection schedules of the equipment, materials, procedures and premises concerned.

Flammability of chemicals, liquids, gases and vapours should be checked out during the safety survey. Readily accessible information should be available through material safety data sheets and Hazchem Codes (see ¶914 and ¶916).

Regulations may provide for the provision of intrinsically safe extraction systems where explosive vapours and dusts result from the work process.

¶924 Hazards associated with high-piled storage

Results of American studies suggest that heat released from a 2.4 metre high stack of burning combustibles was approximately 48,000 British Thermal Units per minute. If the height was doubled to 4.8 metres, however, the increase in heat multiplied nine times to around 440,000 British Thermal Units per minute. In tests of fires involving high-piled storage, roof temperatures have reached almost 900°C (steel loses strength and will have trouble supporting its load at less than half this temperature).

The other problem regarding high-piled storage is that many materials can collapse and so spread the fire to other areas or completely shield the fire from the sprinkler discharge. Depending on the weight of contents, it could of course help to collapse the surrounding walls or structure of the building in which the contents are contained.

The implication of these facts is that special attention should be paid to minimising areas of high storage of flammable materials and evaluating the degree of risks involved. State regulations or ordinances may also set storage restrictions.

¶925 Hazards associated with electronic equipment

The electronic components of computers, automatic storage systems, automatic processing plant and similar equipment are temperature sensitive, and their characteristics can be changed permanently if they are heated to temperatures beyond prescribed limits. These temperatures are lower than those that would render most other items of the plant inoperative. Smoke and corrosive products of combustion can impair the operation of electronic circuits to the point of major failure, necessitating complete replacement of equipment.

Electromechanical equipment can be damaged by minor fires that destroy cable insulation, damage relays and render machinery unworkable because components buckle under thermal expansion. Printed circuit boards can add fuel to a fire.

Any form in which electronic data processing information is stored can be distorted or melted by fire, or made faulty (to the extent that errors occur when used) by lesser heat.

The extent to which the organisation is dependent on the electronic equipment or stored data will indicate the capital outlay that is reasonable in order to provide protection against fire. Where the efficient operation of the organisation is dependent on the equipment and data, its destruction could prove disastrous.

Because of the great importance to an organisation of this type of equipment, it is advisable to seek specialist advice when purchasing and installing it, and considering the building layout in which it is to operate. Matters for consideration include wiring, paper storage, waste disposal, lighting, access, isolation from other risk areas, ventilation, detection and extinguishing equipment.

Some means of back-up record system should also be investigated, so that even if the equipment is destroyed, the organisation can continue to function.

¶926 Warning systems

Warning systems should be planned for in the original building design (or future alteration). If fire or any other alarm or warning devices are installed, employees should be aware of what the devices indicate and what they, the employees, are required to do. This could be done at the induction stage and reinforced by sign-posting on site and trial alerts. Regular testing of alarm systems should be carried out.

There are various types of alarms on the market, which can be operated either manually or automatically. Detection equipment is also available. Smoke/heat detection equipment can be incorporated into a system to act as a self-alarm or to notify the fire brigade.

Relevant government departments can advise on approved types of products, and equipment manufacturers can advise on prices.

¶927 Fire-fighting equipment

With fire extinguishers, it is important to understand the different types in use, and which particular type of fire each should be used for. The table opposite provides an indication.

Work Environments

Fire-fighting equipment

TYPE OF EXTINGUISHER	STANDARD COLOUR	TYPE OF FIRE				
		Class A	Class B	Class E	Class F	
		Wood, textiles, paper, rubbish	Flammable liquids	Live electrical equipment	Cooking oils and fats	Comments
Water	Red	Yes	No	No	No	Dangerous if used on electrical fires
Foam	Stainless steel with blue band / Red with blue band	Yes	Yes	No	Yes	Dangerous if used on electrical fires
Wet chemical	Oatmeal / Red with oatmeal band	Yes	No	No	Yes	Dangerous if used on electrical fires
Carbon dioxide	Red with black band	Yes	Yes	Yes	Yes	Not specially suitable outdoors
Powder	Red with white band (ABE)	Yes	Yes	Yes	No	Can cause serious damage to sensitive equipment
	Red with white band (BE)	No	Yes	Yes	Yes	
Vapourising liquid	Red with yellow band	Yes	Yes	Yes	No	Ventilate well after use

¶928 Overall responsibility for fire prevention and control

In each firm there should be one person with overall responsibility for all aspects of fire prevention from training to fire-fighting. Where this responsibility is divided among several people the efficiency of the fire protection arrangements is diminished.

All firms should therefore have a fire warden, who should be given full managerial backing. Where an organisation is small, a manager should be responsible for fire protection as well as his/her other duties.

One of the first jobs will be to draw up a fire training program for all employees, including temporary, part-time and seasonal. All the points covered in this guide should be included in the program, which must also take into account induction training for new employees.

The same person should have responsibility for handling other types of emergency, such as bomb scares.

¶929 Emergency evacuation plan

A plan for evacuation of the building in the event of an emergency should be formulated. The plan should be widely publicised (for example, on noticeboards and during each employee's induction program). Instructions should be clearly worded (in more than one language where appropriate) and include plans of the building layout showing locations of exits, meeting points and fire-fighting equipment.

The plan should be drawn up according to the organisation's needs and revised each time there are alterations to procedures, processes or premises. The following summary provides an indication of what items need to be included.

Summary emergency evacuation plan

1. Set out the escape plan including:
   (a) designated routes from each work location within the building, dividing staff numbers to use all exits and reduce congestion;
   (b) where to assemble once clear of the building;
   (c) whom to report to once clear of the building; and
   (d) advice not to go back for valuables and not to waste undue time collecting valuables.
2. Prepare diagrams to set out high risk areas such as kitchen, boiler/refrigeration/plant rooms, cleaners' cupboards, storage areas and hazardous processes.
3. Designate duties of individuals, such as checking corridors, using extinguishers, closing doors, and train accordingly. Ensure alternative staff are also identified in case someone is absent on the day.
4. Names and locations of emergency wardens should be listed on noticeboards.
5. Telephone numbers of fire station, police and ambulance should be listed on telephones.
6. On first discovering a fire, bomb, or threatening situation, notify emergency warden, switchboard, or fire brigade/police/ambulance (as required).
7. Encourage people to remain calm and orderly.
8. Feel surfaces of closed doors before opening them (to determine whether fire is present on the other side).
9. Close doors to fire escapes and windows behind you.
10. Carry out a "search and remove" operation floor by floor, including dressing rooms and toilets.
11. Shut down any potentially dangerous units, such as boilers.
12. Do not use lifts.
13. If escaping through a smoke filled area, keep close to the floor (where there is most oxygen) and do not move too quickly.
14. If trapped, go to an outer room where the door can be shut and try to attract attention from a window.
15. Do not jump from a building unless there is no alternative, conditions make it safe to do so, or it is possible to land on a soft surface.

¶929

¶930 Fire-drill

A well organised "disaster plan" will be of little use if employees are not aware of how they should react to a fire warning. Similarly, management will not know whether the plan will work effectively unless it has had a "test run". Even the latter, however, cannot simulate the fear/panic situation caused by hot smoke and fumes.

Employees need to be aware of the evacuation procedure (see ¶929), how to raise an alarm, and any designated role they are willing to accept. Examples of the latter include warden, part of a fire-fighting team, first aid attendant, or a particular duty such as closing a door or checking a corridor. It is advisable to medically clear each volunteer to ensure that they have no physical impairment, such as heart or respiratory problems.

There is no set period at which a fire/emergency-drill should be held. It may vary according to labour turnover rates and organisation size. It is important, however, that management be confident that employees are sufficiently well aware of the procedure so that any emergency will be handled efficiently.

¶931 Fire prevention and control checklist

HAVE YOU CHECKED ... ?

1. *Housekeeping* — waste containers, combustible material containers, storage, handling of flammable liquids and substances, rubbish removal.
2. *Smoking* — enforcement of rules, "no smoking" signs.
3. *Electrical equipment* — wiring, motors, fuses, fuse panels, switch boxes, cleaning.
4. *Fire doors and exits* — door condition, lack of obstructions, kept open, closing devices, sign posting, fire-drills.
5. *Fire extinguishers and hoses* — pressurising, charging, accessibility, servicing, employee training.
6. *Sprinkler systems* — control valves, lack of obstruction, anti-freeze provisions, alarms, inspection of sprinkler heads.
7. *Hydrants* — accessibility, condition, drainage.
8. *Water supplies* — valves, connection, pump condition, tanks filled, gravity tanks.
9. *Automatic alarms and extinguishers* — instructions to employees, condition of equipment.
10. *Evacuation procedures* — practised and known by all employees.

The above list is not intended to be exhaustive but provides an indication of the matters that should receive attention when checking for safety against fire.

Further references:

Fire safety at work, Department of Employment and Industrial Relations, Canberra: AGPS, 1984.

Further information on fire safety can be obtained from the fire brigade in each State, or the Australian Fire Protection Association Ltd.

Chapter 10

An Introduction to Health and Safety Problems — Wellness in the Workplace

Occupational back pain
Scope of the problem .... ¶1001
Identifying and assessing risk areas .... ¶1002
Preventing manual handling injuries .... ¶1003
Regulation of manual handling .... ¶1004
Occupational cancer
Scope of the problem .... ¶1005
Identification .... ¶1006
Prevention and control .... ¶1007
Synthetic mineral fibres .... ¶1008
Skin cancer .... ¶1009
Occupational skin diseases
Skin diseases .... ¶1010
Common causes of occupational dermatitis .... ¶1011
Diagnosis and treatment of dermatitis .... ¶1012
Preventive measures .... ¶1013
Occupational hearing loss
Extent of the problem .... ¶1014
Measurement of hearing loss .... ¶1015
Measurement of noise levels .... ¶1016
Hearing conservation programs .... ¶1017
Legislative provisions and standards .... ¶1018
Occupational overuse syndrome
The nature of occupational overuse syndrome .... ¶1019
Types of workers affected .... ¶1020
Reducing the causes .... ¶1021

Stress

Stress

What is stress .... ¶1022

Causes of stress .... ¶1023

Effects of stress .... ¶1024

Stress and the individual .... ¶1025

Methods of alleviating stress .... ¶1026

OCCUPATIONAL BACK PAIN

¶1001 Scope of the problem

Injuries resulting from the manual handling of loads comprise the largest single category of workplace injuries, typically accounting for about one third of all injuries.

There is a strong association between manual handling and work-related back pain. About 70% of back injuries are strains and sprains resulting from manual handling, though back pain can often be caused by slips, trips and falls. Back pain is the most common type of manual handling injury, though injuries to the shoulders, neck, knees and ankles can also result from manual handling.

Manual handling is defined by the National Occupational Health and Safety Commission's *National Standard for Manual Handling* as "any activity requiring the use of force exerted by a person to lift, lower, push, pull, carry or otherwise move, hold or restrain any animate or inanimate object".

As well as the National Standard, the National Commission has also produced a *National Code of Practice for Manual Handling*. State and Territory legislation is based on this standard and code of practice (see ¶1004).

The term "manual handling injuries" refers to a number of conditions. Sprains, strains, and torn muscles or ligaments are the most common injuries. Injury to spinal discs is another more serious back condition. The effect of the injury and the type and location of the pain vary greatly depending on how far up the back the injury occurs.

Back injuries or other manual handling injuries may be particularly distressing to the worker involved. There is usually no visible sign of the injury, yet it causes a great deal of pain, severely limits the worker's home life, and may drag on for many months. Because there is no visible sign of injury, some tend to be labelled as "malingerers", which further aggravates

Wellness in the Workplace

the feeling of loss by the employee. Twenty per cent of workers with back injuries do not return to work within 60 days, and back injuries account for a third of all long-term compensation claims.

Manual handling injuries are sustained most frequently when lifting or setting down loads. An injury may be caused by a single lift, or as a cumulative effect of years of moving loads. Other situations that may lead to manual handling injuries include:

- carrying, pushing or pulling loads;
- operating levers or other mechanical devices;
- maintaining awkward posture over a long period, or while exerting force; and
- twisting or bending the spine, particularly while exerting force.

Personal factors such as poor levels of fitness, typical sleeping or sitting postures, cold muscles, tension or even the types of physical exercise engaged in can make people more vulnerable to a manual handling injury.

¶1002 Identifying and assessing risk areas

The first step in reducing manual handling injuries at a workplace is to identify those jobs/tasks that could cause such injuries. Work methods and human characteristics also need to be taken into account. The following table lists factors for consideration.

Job/task factors	Work methods	People
• Workplace layout	• Trained lifting	• Health
• Frequency/duration	• Team lifting	• Physical capacity
• Environment	• Work flow	• Skill
• Force (weight)	• Organisation	• Experience
• Nature of load	• Accident reporting	• Age
• Type of handling	• Personal protection	
• Mechanical assistance		

While it has been traditional to concentrate on the weight of an object being handled, that is only one factor in determining whether a particular manual handling task presents a risk. A detailed checklist of risk factors is provided below. It is advisable to consider all the matters referred to when assessing manual handling risks. In summary, the major risk factors are:

- weight of the object (or force applied);
- work posture (distance of the object from the body and position of the body in relation to the object);
- frequency of lifting;
- duration of lifting; and
- preparation for the task, ie warm-up.

The characteristics of the person performing the job/task will also influence the likelihood of injury. Factors such as gender, age, strength, and history of back injury are known to affect the risk of injury.

In the following checklist, a positive answer indicates the possibility of a problem. For example, if there is a "Yes" answer to the question "Is the operator required to bend over frequently at low working heights?", this indicates a risk that may need to be addressed when control strategies are considered.

Manual handling risk factors — Checklist

- Does the operator sit or stand constantly with little opportunity to move around?
- Is the operator required to bend over frequently at low working heights?
- Are extended reaches required by the operator to perform a task? If so, are these tasks sustained for at least one minute continuously or on a repetitive basis?
- Is the operator required to undertake work above shoulder level? If so, is this working posture sustained for at least one minute or on a repetitive basis?
- Is the operator required to undertake work in any of the following awkward postures or conditions:
    - (a) crouching while doing a task for longer than one minute;
    - (b) twisting while undertaking manual handling;
    - (c) handling of large sized loads such as boxes or cases of more than 500 mm (18") in any dimension;
    - (d) handling of sheet materials or other large sized loads without straps, special holders, or a second person;
    - (e) handling of loads that can shift suddenly, such as live animals, liquids or bags of loose material;

¶1002

Wellness in the Workplace

    (f) handling of loads where the load is not shared equally between both hands or lifted single-handed, resulting in imbalanced posture;
    (g) sustained (more than 30 seconds) pushing and pulling of heavy loads;
    (h) pushing or pulling an object across the front of the body;
    (i) handling tasks that require an operator to bend to one side to lift an object or exert a force, resulting in an imbalanced posture;
    (j) inadequate clearances for moving the feet while doing a task that may result in an imbalanced posture or twisting;
    (k) a fixed workplace where it is not possible to adjust the location or heights of equipment being used; and/or
    (l) simultaneous actions, especially with static muscle loading (ie unsupported limbs or application of force without any movement of limbs)?
- For a seated operator, is there sufficient lumbar support (support for the lower back)?
- Is the operator required to undertake occasional lifting or lowering of objects greater than 16 kg (35 lb)?
- Is the operator required to undertake any frequent lifting or lowering of objects greater than 16 kg (35 lb)?
- Is the operator required to lift very heavy objects (ie greater than 55 kg (120 lb))?
- Does the operator lack suitable protective clothing for use during handling operations?
- Are suitable mechanical aids available for use during handling operations? (See section on Mechanical aids at ¶1003.)
- Does the operator have appropriate training for the manual handling task?
- Do any of the following work environment factors make manual handling operations more difficult:
    (a) thermal environment;
    (b) noise levels;
    (c) airborne contaminants including dusts, fumes, gases and vapours;
    (d) lighting glare (direct or indirect); and/or
    (e) vibration levels?
- Does the operator handle objects with protrusions and/or sharp edges without mechanical assistance or appropriate protection?
- Do physical difficulties (eg. excessive fatigue) arise from the time taken and/or distance travelled while undertaking the handling operations?
- Are there occasions where loads, while being handled, obstruct the operator's field of view?
- Do manual handling activities occur in work areas that have unduly rough floor surfaces, or floors that slope significantly or are slippery?
- Have strain or sprain injuries occurred while individuals were engaged in team lifting?

Continued over

¶1002

> **Sitting workplaces**
> - Do operators experience difficulty reaching or handling needed items within the seated work space?
> - Do items being handled require hands to work at an average level of more than 15 cm (6;dp) above the work surface?
> - While handling objects in the seated position, is large force exerted without mechanical aids?
> - Are fine assembly or writing tasks done for a majority of the shift time?
>
> **Standing workplaces**
> - If objects weighing less than 4.5 kg (10 lb) are handled, could the operation be more suited to a varied work posture that allows change between sitting and standing?
> - Are manual handling operations physically separated, requiring frequent movements between work stations?
> - Are downward forces exerted, such as in packing operations? If so, can these forces be eliminated by the use of mechanical aids?
> - Could suitable floor mats help to reduce the discomfort of operators whose job requires them to stand throughout the shift?
>
> **Sit/stand workplaces**
> - Could more variety in posture be introduced to jobs where multiple tasks are performed, some best done sitting and others best done standing?

¶1003 Preventing manual handling injuries

Preventing back injuries and other manual handling injuries means, as far as possible, eliminating the risks involved in manual handling tasks. Risk control can be addressed in three major ways:

- job redesign;
- provision of mechanical aids; and
- training.

Of these three alternatives, job redesign is the preferable option. It can often be used to eliminate manual handling tasks, or to alter them so that they are not stressful on the back.

Job redesign

Whilst job redesign must be considered in the light of the individual job/task, it is possible to provide some examples.

In order to reduce bending, it may be possible to raise the working level. Adjustable height work-benches may solve the problems of different workers using the bench, or different tasks being carried out on the same bench. Materials or tools that will be used at the working level should not be stored at a lower level.

Twisting can be reduced by providing all tools and controls in front of the worker, or by allowing workers sufficient space to move their whole body.

The risks of lifting can be decreased by reducing the size of loads to be lifted, or by increasing the size of loads so that they must be lifted mechanically. The most commonly used materials and the heaviest materials should not be stored at ground level or above shoulder level.

Mechanical aids

Mechanical aids to manual handling tasks include:

- hooks;
- crowbars;
- trolleys;
- belt conveyors or roller conveyors;
- jacks;
- cranes and hoists;
- fork-lifts and other industrial vehicles; and
- adjustable tables/platforms.

Training

Proper training in manual handling tasks should be provided regardless of other preventive strategies adopted. The training should encourage workers to consider better ways of performing tasks (job redesign) as well as covering handling techniques. Workers should be made aware of the risks involved with each job/task, and the way of performing that job/task with least risk.

Where mechanical aids are used, it will be necessary to provide training in the proper use of the equipment.

There has been some debate over the best lifting technique. However, the following principles are generally recognised as important when training people in safe manual handling:

1. *Plan.* The person undertaking the manual handling should assess the load and determine how it will be handled and where it will be placed as one way of avoiding over-exertion injuries. At this stage, the person can assess whether a handling aid or the assistance of another person is needed.
2. *Determine the best technique.* All factors should be taken into account when determining the best technique. The person should do the manual handling efficiently and rhythmically. The best technique involves suitable balance and avoidance of unnecessary bending, twisting and reaching. Above all else, the manual handling technique must be specific to the requirements of the task.

¶1003

3. *Take a secure grip on the object being handled.* The grip helps determine how safe the task will be. Wherever possible, a comfortable power grip (using the whole hand) should be used, rather than a pinch grip (the thumb and fingers only).
4. *Pull the load close to the body.* In lifting in particular, it is important to have the centre of gravity of the load close to the body. This prevents excessive stress on the back, and makes the strongest muscles of the arms available to hold the load. It is important to minimise the effects of acceleration by handling the load slowly, smoothly and without jerking.
5. *Vary handling tasks with lighter work.* The job should be designed to provide alternative tasks that do not heavily stress the same muscles.

¶1004 Regulation of manual handling

Employers need to assess all manual handling risks in their workplace, and then use job redesign, mechanical aids and training to reduce those risks. This is the approach taken by all States and Territories that have manual handling legislation consistent with the National Commission's *National Standard and Code of Practice on Manual Handling*.

Legislation and codes of practice are in place in all jurisdictions:

NSW:

Occupational Health and Safety (Manual Handling) Regulation 1991

Code of Practice for Manual Handling 1991

Vic:

Occupational Health and Safety (Manual Handling) Regulations 1999

Code of Practice for Manual Handling

Code of Practice for Manual Handling in the Furniture Removal Industry

Qld:

Advisory Standard for Manual Tasks 1999

Advisory Standard (Code of Practice) Manual handling — the Handling of People 1992

Advisory Standard (Code of Practice) Manual handling in the Building Industry 1991

SA:

Occupational Health, Safety and Welfare Regulations 1995

Code of Practice for Manual Handling

WA:

Occupational Health and Safety Regulations 1996

Code of Practice for Manual Handling

Code of Practice — Manual Handling in the Building Industry

Code of Practice for Manual Handling — the Handling of People

Tas:

Workplace Health and Safety Regulations 1998

NT:

Work Health (Occupational Health and Safety Regulations) 1992

Approved Code of Practice for Manual Handling

ACT:

Occupational Health and Safety (Manual Handling) Regulations 1998

Code of Practice and Standard on Manual Handling

Further references:

(Note that some of these references are no longer available from the organisations that produced them but may be found in libraries with special OHS collections.)

National Standard for Manual Handling and National Code of Practice for Manual Handling, February 1990 (National Occupational Health and Safety Commission).

Preventing Back Pain at Work, resource kit (National Occupational Health and Safety Commission).

Core Training Elements for the National Standard for Manual Handling, National Occupational Health and Safety Commission, 1995.

BackPak, a Guide to the Manual Handling Regulation, NSW WorkCover Authority, 1995.

Back injuries. Statistical Profile 1994/95/, NSW WorkCover Authority, 1995.

¶1004

Guidance Note for Manual Handling in the Retail Industry, National Occupational Health and Safety Commission, 1992.

Ergonomics for the Control of Sprains and Strains in Mining, B. McPhee, National Occupational Health and Safety Commission/Joint Coal Board, 1994.

Managing Manual Handling Risk — a Resource Kit, Health and Safety Organisation (Vic).

Reducing Manual Handling Injuries, Health and Safety Organisation (Vic).

Back care at Work, WorkSafe Western Australia.

Ergonomic Workstations for Keyboard Operators, WorkSafe Western Australia.

It's Your Back Handle with Care, WorkSafe Western Australia.

Backcare at Work Training Package, WorkSafe Western Australia.

Guide to Manual Handling in the Wholesale and Retail Industry, WorkSafe Western Australia.

Understanding Manual Handling, SA Occupational Health and Safety Commission.

Manual Handling: Health and Safety Issues for Women Workers, SA Occupational Health and Safety Commission.

Manual Handling, Work Health Authority (NT).

OCCUPATIONAL CANCER

¶1005 Scope of the problem

Cancer may be described as a growth or tumour that tends to spread and to reproduce itself. Occupationally induced tumours are no different in type and nature from those arising from non-occupational factors.

Occupational carcinogens, or cancer inducing agents, can include chemical substances, physical agents and work processes.

Estimates of the percentage of work-related cancer vary, and are complicated by the fact that life-style factors (away from the workplace) such as social class, personal habits, nutritional standards and area of residence, may contribute to the occurrence of cancer. An example is skin cancer, which is becoming recognised as job-related in various outdoor work situations, although obviously it can also be brought on by leisure time spent in the sun. However, it is clear that certain substances and work processes are in themselves carcinogenic.

¶1006 Identification

More important than determining the overall effect of occupational cancer is the identification of occupational groups who are at high risk. A good example is provided by asbestos workers. It is now well established that asbestos dust is a source of lung cancer and mesothelioma (cancer of the lining of the lung). However, the late response to the recognition of asbestos dust as a source of cancer has resulted in deaths and disease in all industrial countries.

Identification of cancer-inducing agents is not easy and is basically the concern of research scientists. As work processes change over the years, and as there is almost invariably a long period between initial exposure to a carcinogen and the final onset of disease, a study may not precisely identify the effect of a specific carcinogen used in a particular work process.

The National Occupational Health and Safety Commission's national model regulations list prohibited carcinogenic substances (which are prohibited except for use in research and analysis) and notifiable carcinogenic substances (which require notification of use to the relevant public authority). (See table of *Prohibited Carcinogenic Substances* on the following page.)

PROHIBITED CARCINOGENIC SUBSTANCES

Substance Name (Chemical Abstract Number)

2-Acetylaminofluorene (53-96-3)

Aflatoxins

4-Aminodiphenyl (92-67-1)

Amosite (12172-73-5) (brown asbestos) — except for removal and disposal purposes and situations where amosite occurs naturally and is not used for any new application.

Benzidine (92-87-5) and its salts (including benzidine dihydrochloride (531-85-1))

bis(Chloromethyl) ether (542-88-1)

Chloromethyl methyl ether (107-30-2) (technical grade which contains bis(chloromethyl) ether)

Crocidolite (12001-28-4) (blue asbestos) except for removal and disposal purposes and situations where crocidolite occurs naturally and is not used for any new application.

4-Dimethylaminoazobenzene (60-11-7)

2-Naphthylamine (91-59-8) and its salts

4-Nitrodiphenyl (92-93-3)

NOTIFIABLE CARCINOGENIC SUBSTANCES

Substance Name (Chemical Abstract Number)

Acrylonitrile (107-13-1)

Benzene (71-43-2) — when used as a feedstock containing more than 50% of benzene by volume

Chrysotile (12001-29-5) (white asbestos) — when used for the manufacture of asbestos products.

Cyclophosphamide (50-18-0) (cytotoxic drug) — when used in preparation for therapeutic use in hospitals and oncological treatment facilities, and in manufacturing operations

3,3'-Dichlorobenzidine (91-94-1) and its salts (including 3,3'-Dichlorobenzidine dihydrochloride (612-83-9))

Diethyl sulfate (64-67-5)

Dimethyl sulfate (77-78-1)

Ethylene dibromide (106-93-4) — when used as a fumigant

4,4'-Methylene bis(2-chloroaniline) (101-14-4) — MOCA

2-Propiolactone (57-57-8)

o-Toluidine (95-53-4) and o-Toluidine hydrochloride (636-21-5)

Vinyl chloride monomer (75-01-4)

¶1006

¶1007 Prevention and control

It is vital to determine which substances used at a workplace may be carcinogenic. Material Safety Data Sheets should be consulted to discover the make-up of each substance and any known health effects of that substance. It may be necessary to check a list of chemicals used at the workplace against a list of known carcinogens, such as the National Occupational Health and Safety Commission list provided at ¶1006 above. All areas of the workplace should be considered.

Each carcinogen should be documented regarding measures to be taken if exposure occurs. Workers should be aware of and involved in this process. Affected areas should be designated clearly in some way.

The following checklist of preventive measures should also be observed.

1. Consider substitution with a non-carcinogenic substance wherever possible.
2. Minimise contact with the carcinogen, through work layout, equipment design, repair/maintenance, waste disposal, emergency procedures, washing facilities and decontamination units.
3. Avoid burdening workers with time limits and deadlines during hazardous operations.
4. Operating instructions should be written, expressed in simple and clear terms, and allow for no deviation or improvisation.
5. Where special clothing is necessary, clean replacement clothing should be available, with special provisions for collecting and laundering or disposal of contaminated clothing. Information on hazardous materials should be conveyed to the cleaning sources to avoid risk to their employees. Change-rooms should be arranged so that there is no contact between workclothes and ordinary clothes.
6. Partaking of food and drink, as well as smoking, should be prohibited within the area. Suitable precautions to avoid contamination in lunchrooms, etc, should also be taken.
7. Regular monitoring of equipment (for leakage) and spot sampling of the working environment should be undertaken and measurements recorded.
8. Personal sampling devices should be used regularly to measure specific worker exposure, especially when changes in work processes occur. Monitoring records should be kept. Each employee should be informed of personal results.
9. Workers should be kept constantly informed of the nature of the hazard and all safety precautions initiated to minimise exposure.

Continued over

10. Attention to the selection of employees to work in hazardous situations is important, in terms of technical competence, ability to work in a team, personal hygiene, and ability to give and understand instructions. Use of outside contractors should be avoided, unless they specialise in that particular field, or are able to comply with organisational directives (which should be written into the contract).
11. Company medical/health staff must be informed of the hazard and of any accidents or similar incidents.

Further references:

National Model Regulations for the Control of Scheduled Carcinogenic Substances, National Occupational Health and Safety Commission, 1995.

National Code of Practice for the Control of Scheduled Carcinogenic Substances, National Occupational Health and Safety Commission, 1995.

Guidelines for Laboratory Personnel Working with Carcinogens or Highly Toxic Chemicals, National Health and Medical Research Council, 1990.

¶1008 Synthetic mineral fibres

Synthetic mineral fibres (or man-made mineral fibres) refers to a variety of fibres made from such things as rock, slag, glass and clay. Examples include fibreglass, glasswool, glass filament, mineral wool (or rockwool) and ceramic fibres. These fibres are used for thermal and acoustic insulation and are coming into increasing use, in part as a replacement for asbestos.

Concerns have been expressed that synthetic mineral fibres (SMFs) can be inhaled and lead to lung cancer in the same way as asbestos. While this has not been fully established, SMFs are believed to be less hazardous than asbestos fibres. SMF dust contains fewer respirable fibres than asbestos dust, and SMFs do not break down very readily into smaller and even more easily respirable fibres. Ceramic fibres may be the most hazardous, due to their having smaller diameters than other SMFs.

Employers and workers should minimise exposure to SMFs. The National Occupational Health and Safety Commission has set a national standard of 0.5 respirable fibres per millilitre of air, as the maximum acceptable exposure to SMFs. Note that SMFs can be purchased with binders and de-dusting oils added, which may reduce exposure to SMF dust.

Exposure to SMF dust can be reduced by various engineering and administrative methods such as:

- obtaining SMFs in pre-cut and wrapped form;
- providing localised dust extraction where it is necessary to cut SMFs;

Wellness in the Workplace

- cleaning up waste and packaging regularly; and
- segregating areas where SMFs are in use from other work areas.

Where dust exposure is likely to be significant, employees should wear a respirator approved for SMF dust; long-sleeved, loose fitting clothing and gloves (to prevent skin irritation); and eye-protection and head covering (for overhead work).

Codes of practice

The *National Code of Practice for the Safe Use of Synthetic Mineral Fibres* has been adopted by South Australia, Western Australia, the Northern Territory, the Australian Capital Territory and New South Wales.

¶1009 Skin cancer

Occupational skin cancer is emerging as a serious problem in Australia. Australia has the highest rate of skin cancer in the world, with two out of three Australians who live to the age of 75 developing some type of skin cancer. Obviously it is not possible to attribute all sun damage to outdoor work, with a mixture of work and non-work exposure being responsible in each case. However, outdoor workers should ensure that they are adequately protected against sun damage to minimise their chances of developing skin cancer.

Skin damage from the sun is caused by ultraviolet (UV) radiation. In addition to outdoor workers, there are a number of occupations where workers may be subject to artificial sources of ultraviolet radiation. Note also that ultraviolet radiation passes through water and is reflected very efficiently from snow.

The following list sets out those occupations with potential exposure to ultraviolet light. Where the source of the UV radiation is not sunlight, it is noted.

- Agriculturists.
- Animal farmers.
- Armed personnel.
- Barbers (artificial germicidal UVL).
- Bricklayers.
- Construction workers.
- Cosmeticians (artificial germicidal UVL).
- Council workers.
- Fishermen and women.

- Gardeners.
- Gas, water, sewage and drainage pipeline workers.
- Geologists.
- Greenkeepers.
- Laboratory technicians (artificial germicidal UVL).
- Lifesavers.
- Lithographers (artificial UVL used in drying and curing).
- Loggers and timber workers, tree fellers and similar.
- Maintenance workers (artificial UVL arc welding equipment).
- Oilfield workers.
- Open-cut miners.
- Paint processors (artificial UVL used in drying and curing).
- Pipe cutters (artificial UVL arc welding equipment).
- Pipeline workers (artificial UVL arc welding equipment).
- Plastic workers (some artificial UVL used in drying and curing).
- Police officers.
- Printers (some artificial UVL used in drying and curing).
- Railway maintenance workers.
- Restaurant/kitchen staff (artificial germicidal UVL).
- Road workers.
- Sailors.
- Ski instructors.
- Sportsmen and women.
- Surgeons, nurses, operating room personnel (germicidal UVL).
- Surveyors.
- Unprotected bench workers (adjacent to arc welding).
- Welders (especially those using Tungsten Inert Gas (TIG)).
- Wood curers (some artificial UVL used in drying and curing).

Ultraviolet radiation can have the following deleterious effects on the skin.

Sunburn: Reddening of the skin, blistering, and later peeling of the skin.

Keratoses: (Also known as sunspots.) Dry, rough, stable spots on the skin. These are not cancers but indicate excessive exposure to UV.

Skin cancer: Various forms in Australia. Most dangerous form is malignant melanoma, which appears as a spot or mole that changes colour and shape over months.

Eye damage: UV can cause damage to the retina of the eye and cataracts of the eye.

Preventive measures

There are a number of steps employers can take to reduce the exposure of outdoor workers to sunlight. One essential element amongst these is the education of workers to the dangers that excess exposure to the sun can cause.

Other measures include:

- rescheduling work so that workers are inside between the hours of 10.00 am and 2.00 pm (11.00 am to 3.00 pm during daylight saving) when UV radiation is strongest;
- taking advantage of existing shade or creating shade using canopies and other portable shade structures; and
- supplying and recommending personal protection devices such as hats with wide brims or long flaps, sunscreens, sunglasses and loose protective clothing to cover skin.

Other references:

Guidance Note on the Protection of Workers from Ultraviolet Radiation in Sunlight, National Occupational Health and Safety Commission, 1991.

Protection from Sunlight, National Occupational Health and Safety Commission, 1992.

Skin Cancer and Outdoor Workers — a Guide for Workers, WorkCover NSW.

Skin Cancer and Outdoor Workers — a Guide for Employers, WorkCover NSW.

AS/NZS 4399 Sun protective clothing — evaluation and classification.

OCCUPATIONAL SKIN DISEASES

¶1010 Skin diseases

"Dermatitis" is a general term used to describe any inflammation of the skin. "Contact dermatitis" refers to those conditions that result from direct contact of the skin with an external agent. "Irritant contact dermatitis" is caused by an irritant producing cell damage in the skin with which it comes in contact. Only the skin contacted by the irritating agent becomes affected. "Allergic contact dermatitis", which is less common, is caused when contact with a chemical agent provokes an allergic reaction.

Dermatitis that arises from within the body is known as endogenous. Endogenous skin diseases are rarely caused by occupational factors, although they may be aggravated by agents at the workplace.

Contact dermatitis is characterised by redness, swelling and blistering of the skin, then formation of scales and crusts. After continued irritation the skin becomes thickened and leathery, with spotty areas of pigmentation.

Other skin diseases associated with workplaces

This commentary will concentrate on the various forms of dermatitis, the major occupational skin disease. However, other skin diseases that may occur are described briefly below.

- *Skin keratoses and cancers*: skin damage caused by excessive exposure to the sun (in particular, to ultraviolet radiation). It is vital for Australian workers who spend a large amount of time outside to protect their skin against sun damage. This topic is considered in detail in the section on occupational cancers, at ¶1009.
- *Contact urticaria ("hives")*: a reaction occurring within a few minutes to a couple of hours after exposure of the skin to certain substances. Balsams, alcohols, some antibiotics, nickel, platinum salts and cobalt are some examples of agents which can cause hives.
- *Photosensitisation*: some substances make the skin particularly vulnerable to sunburn. Examples include coal tar pitch and creosote (used as a wood preservative).
- *Occupational acne*: acne can be caused by the irritant effect of mineral oils on hair follicles in the skin. Workers on metal-cutting machines are susceptible. Acne can also be associated with fats and oils used to prepare fried foods.
- *Chloracne*: a far more serious acne-like condition caused by contact with certain chemicals. The condition lasts for many years after exposure has

Wellness in the Workplace

ceased. Some dioxins (notably 2,3,7,8 tetrachloro-di-benzodioxin (TCDD)) are known to cause chloracne.

- *Scleroderma*: roughening and thickening of the skin. Agents known to induce scleroderma include silica dust, vinyl chloride monomer, trichloroethylene and tetrachloroethylene (perchloroethylene).
- *Folliculitis*: this refers to sores arising from irritation of hair follicles. Cutting oil used in association with metal-cutting equipment is the major cause.
- *Low humidity occupational dermatitis*: it is now recognised that particularly dry air can lead to rashes and other skin irritations. This is caused by the outer layer of the skin losing part of its water content. Affected skin also becomes more susceptible to irritants and allergens, that can cause contact dermatitis. Dry air is found in many modern indoor working environments. Those who work with furnaces and ovens and those who work in computer rooms may be particularly susceptible.
- *Perionychia*: this refers to inflammation around the fingernails, found in workers who frequently have wet hands. Bar workers, for example, are susceptible.

¶1011 Common causes of occupational dermatitis

The following list sets out some of the major agents known to cause occupational dermatitis, along with the industries in which those agents are found. The list is taken from *Health and Safety Bulletin, No 50, June 1987*, published by the ACTU/VTHC Occupational Health and Safety Unit, but is not intended to be exhaustive.

- cement — construction workers, cement manufacture workers, cement products;
- epoxy resins — construction workers, applicators in light industry;
- acrylic plastics — moulded articles producers, sealant users, printers;
- surfactants, soaps, detergents — kitchen and hotel workers, hospital workers, metal industry;
- mineral oils — metal machining workers;
- solvents — degreasing and cleaning workers;
- tar, pitch — roofing and bitumen workers, and handlers of creosote-treated wood, eg railway sleepers, electricity poles;
- soldering fluxes — electronics workers;
- pesticides and preservatives — manufacturing workers and applicators in farms, forestry, and users of sprayed products;

- printing inks and plates — printing workers;
- pharmaceuticals — manufacturing workers;
- hairdressing preparations — salon workers; and
- film processing chemicals — film laboratory workers.

Cement is the single biggest cause of occupational dermatitis. The alkalinity of cement causes irritant contact dermatitis, while the chromate content can cause allergic contact dermatitis. It is recommended that workers handling cement, particularly wet cement, be provided with protective clothing to avoid skin contact. Ferrous sulphate can be added to cement products to prevent chromate dermatitis, although it is better if this is done by purchasing cement additives which already contain ferrous sulphate.

¶1012 Diagnosis and treatment of dermatitis

Effective treatment of occupational dermatitis depends primarily upon an accurate diagnosis. A correct diagnosis can lead to the elimination/control of the noxious agent and thus to the prevention of similar problems in both the patient and other workers. At the same time, the appropriate treatment can be provided.

Because of the importance of correct diagnosis, all patients with suspected occupational dermatitis should be referred to a dermatologist. Even experienced clinicians may find it difficult to distinguish between dermatitis and other conditions, such as tinea.

The taking of a thorough history is essential to the correct diagnosis of dermatitis. This should include exposure to any known or suspected irritants or allergens. The possibility of irritants/allergens away from the workplace should also be considered. Any improvement or lack of improvement on weekends or holidays can be an important clue as to the existence of irritants/allergens at the workplace.

Suspected allergic contact dermatitis can be diagnosed through patch testing (note that patch testing is not appropriate for irritant contact dermatitis). Patch testing involves applying small amounts of suspected chemicals to the skin for 48 hours. There is a standard patch test battery, containing the most common contact allergens, as well as special batteries available for certain industries. Alternatively the worker may supply suspected chemicals or materials from the workplace for use in a patch test. Patch testing should only be performed and interpreted by a dermatologist.

Once a correct diagnosis has been made, removal from the causative agent and the appropriate treatment should result in cure. Employers may

Wellness in the Workplace

need to take the following steps to remove the affected employee from contact with the irritant/allergen:

- providing protective clothing;
- moving the affected worker to a new area within the workplace; or
- replacing the offending agent with an alternative (for example, a particular brand of printing ink containing a known allergen may be able to be replaced with another brand free from the allergen).

In a small number of cases the only way for a worker to avoid the noxious agent will be to change his/her profession.

¶1013 Preventive measures

The National Occupational Health and Safety Commission, in its *Guide to Occupational Diseases of the Skin*, recommends a number of measures that can help to prevent the occurrence of occupational skin diseases. These are summarised below.

1. Where possible the hazard should be eliminated. If this is not possible, a less hazardous chemical or process should be substituted.
2. Engineering controls should be used to minimise exposure to dust, gases, fumes, fogs, mists and vapours. These controls should aim to prevent direct contact of the dust or solution with the skin, as well as avoiding build up of materials on surfaces that may come into contact with the skin.
3. A high standard of hygiene should be maintained in the work environment.
4. A high standard of personal hygiene should also be applied. If necessary, employees should have access to clean work clothes, protective clothing, laundry facilities, showers and change-rooms.
5. Employees should be trained in the correct use and care of protective clothing and equipment.
6. Material Safety Data Sheets should be consulted for details of skin hazards in relation to new and existing compounds.

Further references:

Health and Safety Bulletin, No 49 and 50, "Guidelines on Occupational Skin Diseases", May and June 1987 (ACTU/VTHC Occupational Health and Safety Unit).

Occupational Diseases of the Skin, Guide, November 1990 (National Occupational Health and Safety Commission).

Occupational Skin Diseases, RM Adams (Ed), 2nd ed, WB Saunders, Philadelphia, 1990.

The Journal of Occupational Health and Safety — Australia and New Zealand, Vol 2, No 4, August 1986 (on occupational skin diseases) (CCH Australia Limited).

OCCUPATIONAL HEARING LOSS

¶1014 Extent of the problem

Occupational hearing loss, or "industrial deafness", has become an increasingly significant issue in recent years. The effects are likely to be gradual and, coupled with loss of hearing ability due to advancing age, can be substantial.

It is estimated that between 250,000 and 500,000 Australian employees work in hazardous noise environments (DND 0.33 or more). The National Occupational Health and Safety Commission has stated that occupational hearing loss leads to around 10,000 compensation claims each year, making it the single largest cause of claims.

While leisure activities can also contribute to hearing loss, the extent of this can be overemphasised. Away from the workplace, exposure to loud noise tends to be infrequent and for short periods, though the use of walkman tape players at loud volume, loud audio systems in motor vehicles and frequent attendance at venues where very loud music is played can damage hearing.

The consequences of hearing loss can be extremely disturbing to the individual. The reduction in communication ability is the most serious, although there are all sorts of other domestic noises which we usually take for granted (television and radio, hearing the kettle boil, a car horn signalling warning, and so on). Noise-induced hearing loss often causes people to become resentful and places additional strains on relationships due to the problems in communication. There may be a marked reluctance for a person to admit that they have suffered some hearing loss, and consequently to seek assistance.

At work a noise-induced hearing loss can lead to an inability to hear instructions clearly with consequent poor relationships with fellow workers and immediate supervisors. It may lead to misinterpretation of instructions, and could possibly contribute to accidents and loss of continuity in the work. Telephone conversations against the background noise of a workshop become virtually impossible.

¶1014

¶1015 Measurement of hearing loss

A possible noise hazard and the associated hearing loss is indicated informally by such factors as

- difficulty in communication by speech at the workplace (having to raise the voice at a distance of 1 metre or less);
- ringing in the ears (tinnitus) or difficulty in hearing after exposure to noise;
- temporary loss of hearing or shouted conversation at the end of the workday; and
- an unusual prevalence of deafness in employees.

Informal measurement is not conclusive, however. A noise survey (involving measurement of noise levels) should be undertaken in all areas where hazardous exposure to noise is known or suspected to occur, because an individual will usually only become aware of a change in hearing after damage has occurred. The measurement of noise levels is discussed further at ¶1016

The measurement of a person's hearing ability is known as audiometry. It should only be carried out by a person trained in that skill. An audiometric test generally involves sitting in an insulated room and listening on headphones to sounds of differing tone and loudness. While it is possible to distinguish between noise-induced hearing loss and hearing loss caused by disease, it is not possible to distinguish between noise-induced hearing loss and loss brought on by old age. Nor is it possible to determine which noises were responsible for any loss.

Where the workplace involves a reasonable amount of noise, it is wise to test the hearing acuity of all new employees before they start work. This is:

- for their own information;
- to ensure that in any future compensation claim the employer will not pay for a hearing loss that was not caused at that workplace; and
- to create awareness of a workplace hearing conservation program.

After this tests could be carried out every two years. Note that in some States there are legislative requirements prescribing regular audiometric testing.

¶1016 Measurement of noise levels

The unit of noise is the dB(A) (decibels on the A scale). Decibels are measured on a logarithmic scale, so the 80 dB(A) is 10 times 70 dB(A) and 90 dB(A) is 100 times 70 dB(A). Sound levels for some common noises are provided below.

Sound level in dB(A)	Sound source
140	Jet engine
130	**THRESHOLD OF PAIN**
120	Propeller plane
110	Chain saw
100	Sheet-metal workshop
90	Heavy truck
80	Heavily trafficked street
70	Car
60	Normal conversation
50	Low conversation
40	Quiet radio music
30	Whispering
20	Quiet urban apartment
10	Rustling leaves
0	**HEARING THRESHOLD**

Because hearing loss is a function of the intensity of noise and length of exposure, a measurement called the Daily Noise Dose (DND) has been developed. A DND of 1.00 is equivalent to an exposure of 90 dB(A) for eight hours. Because the dB(A) system is measured along a logarithmic scale, exposure should be halved for every 3 dB(A) increase, as shown in the table on the following page. In general, the legal requirement throughout Australia for most of the 1980s was a DND of 1.00, with hearing conservation action to be initiated whenever workers were exposed to a higher level. This level of exposure was recommended by the National Health and Medical Research Council in 1973 as a reasonable social/economic compromise, but even at this level 28% of workers will experience some significant hearing loss over a working life. For this reason most States have now moved to a maximum DND of 0.33 (equivalent to an exposure of 85 dB(A) for eight hours).

Wellness in the Workplace

Time of exposure to reach different levels of daily noise dose			
For the DND not to exceed			
1.00		0.33	
Average noise level in dB(A)	Hours of exposure each day	Average noise level in dB(A)	Hours of exposure each day
90	8	85	8
93	4	88	4
96	2	91	2
99	1	94	1
102	1/2	97	1/2
105	1/4	100	1/4

Source: Derived from Australian Standard AS 1269

Noise measuring equipment

There are two types of instruments used in the workplace to measure noise:

- a noise level meter; and
- a noise dose meter.

Noise level meters are used to determine the level of noise at a particular place. Some instruments can measure sound levels at different frequencies.

Noise dose meters can be worn by workers, with the microphone as close to their ear as possible, for all or part of the working day. The meter will integrate the noise recorded and give the DND as a direct reading. These are an excellent way to check the effectiveness of a workplace hearing conservation program.

¶1017 Hearing conservation programs

The basic steps for conserving hearing are set out in Australian Standard *AS1269:1998 — Occupational Noise Management*. Briefly they involve the measurement of noise, its reduction where feasible, the use of hearing protection if reduction is not feasible, and audiometry to monitor overall progress. Further essential ingredients include the motivation of workers, allocation of specific responsibilities to supervising staff and involvement of those concerned with safety and/or health in the workplace.

Engineering controls

The removal of noise by engineering methods should be given top priority in any hearing conservation program. Noise specifications should be included in a company's purchasing criteria, and all new equipment should be checked before purchase for the noise that it will produce. Installation of machinery should also be done with noise control in mind.

There is a wide range of steps that can be taken to reduce the noise caused by machinery that is already in place. These include:

- replacing noisy machines with quieter ones;
- placing machinery in sound-proofed rooms;
- placing machinery on vibration-dampening mounts;
- using exhaust-dampening equipment; and
- treating walls to reduce reflected noise.

Administrative controls

Some administrative measures may be introduced to minimise workers' exposure to noise levels. These could include the rescheduling of noisy processes to periods of minimal occupancy, job rotation, and the signposting of noisy areas. Often administrative measures can be used as a temporary solution until engineering solutions can be implemented.

Personal hearing protection

In some cases, elimination of harmful noise is not feasible so that hearing protection will be necessary for the foreseeable future. Fortunately, the Australian method for testing hearing protectors and for rating their performance with a single number has received wide acceptance because it is conservative and relates closely to how such items are used in practice. However, personal protection is never a really satisfactory method of risk control, for reasons such as the following:

- it is not possible to ensure that hearing protectors fit all the time, so they cannot be regarded as giving full protection all the time;
- hearing protection may be uncomfortable to wear, particularly in hot and humid weather;
- ear-muffs become damaged and worn — a system of regular maintenance must be introduced; and
- ear infections are possible through the fitting of plugs when proper personal hygiene is not observed.

¶1017

Introducing a program

A hearing conservation program should be introduced with the full support of management, employees and technically competent persons. The co-operation of employees should be sought in suggesting and designing control measures. An education program can accompany the introduction of the noise conservation program. This should be aimed at all employees, and should cover the effects of noise, its measurement and the use of hearing devices.

The fact that each year sees quieter equipment coming on to the market and that hearing protectors compatible with a wide range of other personal protective equipment are now available negates any excuse for delaying the introduction of hearing conservation programs in the majority of small organisations, as well as in large industries.

¶1018 Legislative provisions and standards

Noise levels and hearing conservation are covered by Regulations in each State and Territory. The exposure standards that have been adopted in all jurisdictions are:

- an 8-hour equivalent continuous sound pressure level (L aeq.8h) of 85 dB(A), referenced to 20 micropascals; and
- a linear (unweighted) peak sound pressure level (L peak) of 140 dB(lin).

National standards

The National Occupational Health and Safety Commission has established a national standard for exposure to occupational noise of an 8-hour noise equivalent of 85dB(A), with a maximum peak noise level of 140 dB(lin). All States and Territories have adopted this exposure standard in legislation.

AS/NZS 1269:1998 Occupational Noise Management is a guide to the technical aspects of noise measurement and assessment, noise control methods, audiometry and the selection of hearing protectors.

Regulations and Codes of practice

NSW:

Occupational Health and Safety (Noise) Regulation 1996

Code of Practice: Noise Management and Protection of Hearing at Work

Vic:

Occupational Health and Safety (Noise) Regulations 1992

Code of Practice for Noise

Qld:

Workplace Health and Safety Regulation 1997

Code of Practice for Noise Management at Work

SA:

Occupational Health, Safety and Welfare Regulations 1995

WA:

Occupational Health and Safety Regulations 1996

Code of Practice for Noise Control in the Workplace

Control of Noise in the Music Entertainment Industry

Tas:

Workplace Health and Safety Regulations 1998

NT:

Work Health (Occupational Health and Safety) Regulations 1992

Adopted national code

ACT:

Adopted national standard and code

Further references:

Noise Control at Work, ACTU Occupational Health and Safety Unit, 1987.

Noise Management at Work Resource Package, a kit containing a control guide publication, a video, fact sheets, a poster and stickers, produced by the National Occupational Health and Safety Commission.

Occupational Noise: National Standard for Occupational Noise; National Code of Practice for Noise Management and Protection of Hearing at Work, National Occupational Health and Safety Commission, 1993.

Core Training Elements for the National Standard for Occupational Noise, National Occupational Health and Safety Commission, 1995.

Occupational Noise Induced Hearing Loss: Prevention and Rehabilitation, National Occupational Health and Safety Commission, 1991.

Noise Management at Work: Control Guide, National Occupational Health and Safety Commission, 1995.

Noise Management: A Trainer's Guide, ACCI, 1990.

Deafness. Statistical Profile, 1993-94, NSW WorkCover Authority, 1996.

¶1018

Do I have a Noise Problem? — Noise Regulation, WorkCover Authority of NSW.

Noise at work, WorkCover Authority of NSW.

Noise Regulation — Hospitality and Entertainment Industry, WorkCover Authority of NSW.

Noise at Work, NSW Health Department.

SHARE manual of documented solutions to noise, HSO (Vic).

Noise at Work, Department of Occupational Health, Safety and Welfare (WA).

Managing Noise at Work, Department of Employment, Vocational Education, Training and Industrial Relations (Qld).

Occupational Noise, Northern Territory Work Health Authority.

OCCUPATIONAL OVERUSE SYNDROME

¶1019 The nature of occupational overuse syndrome

Occupational overuse syndrome (OOS also known as repetitive strain injury) is a term used to cover a variety of soft tissue injuries affecting various parts of the body. These conditions are attributed to various types of work and recreation where overload of soft tissues (such as muscles and tendons) occurs through rapid, repetitive and/or forceful movements, and/ or where static muscle loading is maintained over a period of time. Injury is experienced as pain and weakness that may lead to impaired function of the muscles and other soft tissues.

Occupational overuse syndrome affects various parts of the body, including the neck, shoulders, elbow joint, hands, fingers and wrists. More rarely, people using their feet to operate machinery may suffer injuries in the lower parts of their legs. Probably the best known type of injury is tenosynovitis, which is a strain injury of a tendon or group of tendons about the wrist, causing inflammation of the tendon sheath. Its main symptoms are pain and swelling over the affected tendons, increasing restriction in the use of the hand, progressive weakening in grip strength and a creaking or crunching sensation felt in the affected area upon movement of the thumb.

Causes

According to the National Health and Medical Research Council's Occupational Health Guide on repetition strain injuries, the following factors additional to repetition and muscle strength can contribute towards injuries:

- inefficient methods of work (unsuitable hand movements, unnecessary use of muscles, unnecessarily forceful action);
- postures of the trunk, arm or hand requiring joints to deviate repeatedly or for long periods from the position of normal function;
- static loading of muscles;
- unaccustomed work (beginners and workers who resume work after absence, change of work patterns);
- prolonged repetitive work; and
- too short a cycle time, eg when data processing operators work at speeds of up to 16,000 keystrokes each hour.

Work organisation can also be responsible to some degree for the onset of OOS. Boredom, shift work, poor communication and lack of control over a job may all contribute. Stress may also be a factor. Studies indicate that a greater variety of work, regulating work flow to avoid rush periods, and job enrichment can help reduce the level of OOS (see ¶1105 on reducing the causes of stress).

The syndrome is not only caused by work in a work setting. The combination of the type of work performed and the nature of recreational activities or domestic tasks may be a causative factor for some individuals. Some individuals may be predisposed to the development of this condition, although this theory has not yet been proven.

¶1020 Types of workers affected

As research into this area broadens, the range of occupations considered vulnerable to occupational overuse injuries has widened. Whereas initial studies concentrated mainly upon factory and manual work, a number of white-collar/clerical areas have now been identified as prone.

The National Occupational Health and Safety Commission code of practice provides the following occupations where occupational overuse injuries have been noted to occur:

Process workers;

Machinists;

Cleaners;

Kitchen workers;

Keyboard operators;

Clerks;

Musicians;

Carpet layers;

Hairdressers; and

Mail sorters.

This list is not exhaustive.

¶1021 Reducing the causes

The National Occupational Health and Safety Commission has released a *National Code of Practice for the Prevention of Occupational Overuse Syndrome*. New South Wales, the Australian Capital Territory, and the Northern Territory have adopted this code. Victoria has developed its own code of practice. The National Code states that to:

> "identify and address the causes of OOS it is necessary to consider many aspects of the workplace. The overall conditions of work, including the industrial relations climate, will influence the effectiveness of changes introduced to prevent OOS."

Elimination of possible causes of occupational overuse syndrome will involve attention to job design, work layout and ergonomic aspects such as the following:

- analysis of equipment, tools and working posture to reduce physically stressful movements and postures and allow a variety of movements and postures within each task to rest muscle groups;
- dimensions and positioning of aspects of work layout in relation to the worker's strength and size;
- equipment design, such as handles, grip sizes, provision for adjustment, suspension of heavy items;
- training and instruction of employees (both when commencing and resuming work after suffering overuse injury) in recommended work methods, efficient use of muscles and gradual introduction to repetitive manual tasks;
- job rotation so that workers perform different tasks; and
- regular rest breaks from constant and rapid repetitive tasks, as well as encouraging workers to vary the rhythm of hand movements when performing these tasks. In some cases, where it is not possible for keyboard operators to perform other tasks, agreements have been negotiated that provide for work pauses of 10-15 minutes in each hour. The introduction of pause gymnastics has been useful in some workplaces.

Proper education of workers subject to occupational overuse injuries, and their supervisors, is also an important part of prevention. Education should ensure that employees and supervisors understand the potential health problems associated with their workplace and how to avoid those problems. For example, keyboard operators should be informed about the correct use of furniture and equipment and the importance of correct posture, movement and exercise. Employees may also be informed of the relevance of domestic/leisure activities and their potential for worsening OOS.

Occupational overuse syndrome should not be seen as a purely medical or mechanistic phenomenon. In its consideration of the syndrome, the Australasian Faculty of Occupational Medicine concluded that "it seems many claims of RSI [as OOS was known in the 1980s] were in fact people's communications of discontent or alienation in their employment. Seen in this light, it is not surprising that casting the problem in medical terms and advising people to 'see your doctor if pain persists' did not produce the solution. Better results were achieved when managers were forced to confront the issue and properly address workers' complaints."

Overall, therefore, reducing the causes of occupational overuse syndrome may not be a straightforward matter, and may require the advice and assistance of experts in the field, as may the rehabilitation of injured workers. There is no point in embarking on partial interventions such as only buying ergonomic furniture. Causes are multiple and the solution is complex. Partial attempts at rectification/correction will lead to an unsatisfactory result.

Further references:

National Code of Practice for the Prevention of Occupational Overuse Syndrome, National Occupational Health and Safety Commission, 1994.

Guidance Note for the Prevention of Occupational Overuse Syndrome in Keyboard Employment, National Occupational Health and Safety Commission, January 1989.

Guidance Note for the Prevention of Occupational Overuse Syndrome in Manufacturing Industry, National Occupational Health and Safety Commission, 1992.

Pause Gymnastics: Improving Comfort and Health at Work. A Gore and D Tasker; A book of exercise programs designed for performance during short work breaks. The exercises are specifically devised to prevent, reduce and eliminate discomfort and pain associated with musculo-skeletal injuries such as OOS (CCH Australia Limited).

¶1021

RSI Retrospective — Consensus Statement, Australasian Faculty of Occupational Medicine, 1995.

STRESS

¶1022 What is stress?

In simple terms, stress could be defined in the context of occupational health as the result of the body's reaction to excessive levels of pressure caused by various environmental factors. The level of reaction is strongly influenced by the person's own perception of the stress-causing situation, this perception in turn being a function of personality. Stress could therefore be seen as a set of physical, mental and biochemical conditions within a person's body that reflect the body's attempt to adjust to pressuring situations. In the work context, pressuring situations are those aspects of a job that cause problems for the employee and that require some form of adaptation to their demands.

As these situations occur, certain mechanisms in the body are activated, such as glands releasing hormones or changes in the central nervous system, thus enabling the body to adapt. If exposure to the situations is prolonged, these mechanisms will wear down in time until they become inadequate, and symptoms of excessive stress will develop. It is the continuing nature of demands which leads to stress.

Thus, a stressful work environment can impair general health and well-being. It is recognised that there are certain conditions, such as alcoholism, unstable mental states and certain diseases (see ¶1024), in which stress may be inferred as a component either allowing a vulnerability or exacerbating an existing state. Also, stress can be seen to give rise to injury more directly where, for example, safety precautions are ignored to meet a perceived "pressure" of deadlines or quotas.

This information implies that stress is always a bad thing, which is incorrect, as everyone has an optimal level of stress at which they perform most effectively. A degree of stress can be pleasant and beneficial, and it could be said that people require a certain amount of stress for their lives to be meaningful. The problems begin when the optimal level is exceeded. (Because of these difficulties with the stress concept, some researchers take "stress" as being a term to simply describe a body's adaptive response, then differentiate between "distress" where the adaptive response is perceived to be harmful and "eustress" as in euphoria where the response is at a level of pleasant stimulation.)

Early studies of stress concentrated upon its effects at management and executive levels, but it is now apparent that stress can occur at all employee levels within an organisation.

¶1023 Causes of stress

The range of potential stress-causing situations to be considered at work is very wide, but can be divided into three basic categories, as shown in the following table:

1. Job-induced	
• work overload • role conflict • lack of control over work environment and decision making (affects both blue and white-collar workers) • excessive work pace or deadlines	• work underload • role ambiguity • unsatisfactory feedback • insecurity • continual concentration upon goals; overcompetitive work environment
2. Organisation-induced	
• working conditions (noise, light, poor ergonomics, lack of privacy, vibration, overcrowding) • office politics and idle talk • inappropriate management styles • unsafe conditions • unsatisfactory work equipment • various communication problems	• lack of management support • lack of peer group support • discriminatory practices • sexual harassment • unsuitable staffing levels • lack of training
3. Change-induced	
• unfamiliarity with new job	• changes in plans, procedures, work layout and technologies • new supervisor • new co-workers/subordinates • expectations of change, such as transfer, retirement, retrenchment.

In addition, stress can be induced by domestic and personal problems which are "brought to work" and may interfere with the employee's ability to work satisfactorily by causing conflict of needs and priorities.

Note also that the causes of stress can flow downwards from superiors, upwards from subordinates or laterally from peers (such as office politics).

¶1024 Effects of stress

Employee stress can have drastic effects on performance, productivity and cost levels because the affected employees are prevented from efficiently carrying out their work. Apart from the unfortunate effects on the individual, stress can cause adverse reactions throughout a workplace, such as effects upon employee relations, work group attitudes and disruption to the general work flow.

For the individual, there are three main types of response to stress:

1. psychological changes;
2. behavioural changes; and
3. forms of disease or illness.

Psychological changes that may be induced by stress include feelings of boredom, guilt, pressure, anxiety, tension, irritation, worry and pessimism.

Among the many types of behavioural changes that may provide "clues" to stress are absenteeism, cynicism, deteriorating work performance, "addiction" to work, loss of interest and effort in work, withdrawal, conflict/ abrasiveness, and excessive use of coffee, nicotine, alcohol or medicines.

Stress is able to cause physiological changes in the body, and it is possible to measure such changes by measuring the release of certain hormones. For instance, acute stress may lead to the release of cortisol, adrenaline and noradrenaline from the adrenal glands, which can be measured to determine whether they are above usual levels. These physical changes may be associated with various illnesses, including coronary/heart diseases, hypertension, ulcers, migraines, psychosis, neurosis, depression, skin disorders, insomnia and diabetes.

¶1025 Stress and the individual

An assessment of the causes and effects of employee stress should bear in mind that individual responses to stress-inducing situations will vary widely. Some people are able to tolerate a great deal more stress than others, because they do not react as adversely towards it. Also, the same individual's response will vary over time so that he/she may be more or less vulnerable to the adverse effects of stress.

The difference between individuals can be attributed to many factors personality type, age, education, ambitions/expectations, social background, marital status, level of health/fitness before commencing a job, tendency to smoke, consume alcohol and so on. The effect of factors such as these should be considered when examining the methodology of various studies of stress at work. They can be regarded as factors which are brought to the job.

While it is true that certain individuals will be more susceptible to the effects of stress, this should not be used as an excuse for failing to address workplace stressors. All workers will respond positively to the alleviation of an overly stressful situation.

¶1026 Methods of alleviating stress

Role of the employer

The main responsibility of the employer is to minimise the opportunities for stress to develop. This affects several aspects of human resource management - job design, job satisfaction, working conditions, participation in decisions, clear-cut and equitable policies, training and selection of employees, adequate pay rates, and so on.

Training people to deal with stress

From an individual viewpoint, employees may be able to cope with stress better if they can make use of techniques to relax or seek some other means of coping with the situation. It may therefore be beneficial for management to provide some form of "stress education" to employees, in forms such as seminars, films, printed information or individual counselling.

Stress management courses may help, but by far the best approach is to remove or reduce as many of the causes of stress at the workplace as possible. When introducing such a course, management should demonstrate that they are not trying to place the responsibility for stress solely on the employees.

Treating stress

There are several ways of reducing stress at the workplace. They can be divided into the following broad categories.

1. Attention to the *work environment* (such as job redesign and autonomous work groups).
2. Improving the *interpersonal environment* (for example, by meetings, more positive feedback, team-building approaches and better communication).
3. Assisting the *person* — counselling, psychotherapy, rehabilitation, psychological techniques (such as transcendental meditation, transactional analysis, biofeedback, encounter groups) and assistance with maintaining health and physical fitness (such as providing gymnasium facilities or encouraging health awareness).

Improving communication, especially in terms of allowing employees a say in decision making, is also likely to help by reducing uncertainty and fears. For example, the introduction of new technology to a workplace without discussion and adequate preparation may cause fears of retrenchment. Adequate communication would remove this particular stressor.

Further references:

Health and Safety Bulletin, No 54 and 55, May 1988 and June 1988, "The Process of Stress" (including guidelines on controlling stress at the workplace), ACTU/VTHC Occupational Health and Safety Unit.

A Report on the *1997 ACTU National Survey on Stress at Work*, ACTU 1998.

Stress, the Workplace and the Individual, NSW WorkCover Authority, 1996.

Chapter 11

An Introduction to Health and Safety Problems — Social Issues

Shift work and night work	
Types of shift work	¶1101
Health and biological effects	¶1102
Social effects	¶1103
Stand-by periods	¶1104
Reducing the ill effects	¶1105
Alcohol and drug dependence	
Extent of the problem	¶1106
Detecting the problem	¶1107
Role of the employer	¶1108
Employee Assistance Programs	¶1109
Who is involved in counselling?	¶1110
Violence at work	
What is violence?	¶1111
Incidence of violence	¶1112
Likely targets	¶1113
Risk management approach	¶1114
Smoking and the workplace	
Passive smoking	¶1115
Some solutions	¶1116
Age and experience of workers	
Issues relating to youth/inexperience or age of workers	¶1117
Employer concerns	¶1118
Managing risks for young/inexperienced and older workers	¶1119

SHIFT WORK AND NIGHT WORK

¶1101 Types of shift work

Shift work may take the form of a regular, fixed schedule (such as a night watchman working the same shift every night) or may occur as part of a rotating shift system. It may also occur when people work a number of casual jobs in order to enhance their income.

In rotating shift systems, the most common periods of rotation are long periods (20 days or more) or weekly rotations. Some establishments are now moving to short rotations, with only two or three days on each shift. Twelve-hour shifts are becoming increasingly common, as they seem to involve less disruption to a shift worker's life. Twelve-hour shifts allow for two days (or one) on day shift, then two days (or one) on night shift, followed by four days (or two) off.

Generally, night work offers better financial benefits than day work, and some people find it offers them social benefits as well.

Stand-by periods are another form of working outside usual hours. In some professions (notably maintenance/repair workers and emergency workers) employees may spend a large amount of their spare time "on call". The problems associated with stand-by periods are discussed at ¶1104.

¶1102 Health and biological effects

There is some controversy as to how serious the effects of continuous night and/or shiftwork are. While some studies have claimed that significant adverse effects on employee health may occur over a period of time, others have concluded that there is little correlation between night/shiftwork and deterioration in health. Self-selection among shift workers and individual perceptions of a situation may be significant factors.

As with stress, the effects are likely to depend very much on the individual, but some guidelines can be drawn up to indicate the potential harmful effects of shiftwork or night work.

Firstly, night work involves a reversal of normal biological peaks and troughs (such as brain function, respiration) that are usually at their most active during the day, but are now called on for greater activity at night.

It is possible to test the body's adaptation to night work by measuring the level of certain hormones that are produced in greater quantities during the day. This research suggests that workers never really adapt to working night shifts. The main reason for this appears to be that shift workers invariably revert to "normal" sleeping patterns during weekends, breaks and holidays.

As a result of this, fatigue levels are increased by both the extra exertion needed to work during the "deactivation" period and having less chance to recuperate through sleeping during the "active" period of daytime. Sleep during the day tends to be less deep, more fragmented and shorter by an average of one to two hours. Thus shift work can lead to excessive fatigue, which in turn may cause exhaustion or psychosomatic illnesses. At the workplace, fatigue will adversely affect productivity (through impaired concentration or the need for rest pauses), absenteeism and safety. Fatigue can have consequences after work as well, as it can be a factor in drivers falling asleep at the wheel.

Night work tends to lead to a change in eating patterns, and may cause digestive disorders (ulcers and intestinal problems). Night workers may increase their consumption of fast foods, at the expense of proper meals. Consumption of coffee and "stay awake" drugs may increase. Further, since the body never really adjusts to night work, shift workers may not feel hungry when they finish their work, and may lose a regular eating pattern. Some shift workers have reported a general loss of appetite.

Finally, studies have suggested that workers on rotating shifts have a higher percentage of nervous disorders than those on day shifts. It is possible that sleep problems appear to be the symptom or cause of a more basic metabolic disorder.

Comparison of different shift rotations

Previously it had been assumed that long shift rotations were best for health, as they allowed the worker to adjust to each shift. There is now doubt about this, particularly since workers tend to revert to "normal" sleeping patterns on weekends and breaks. New thinking suggests that rapid shift rotation, with a maximum of three nights on night shift, is a better option. The worker is able to cope with a few consecutive nights of work without too much disruption to bodily systems or social structures. It appears that weekly rotation is the worst of all systems.

Recent studies suggest that employees working a 12-hour shift system have very few of the problems associated with shift work. The employees are able to maintain relatively normal sleeping and eating patterns, and have a large amount of time for family and social life. Typically, 12-hour shift workers have to cope with two nights of work every eight days, which are followed by a large period of rest. However, where the work involves a high degree of application, 12-hour shifts may not be appropriate.

¶1102

¶1103 Social effects

Night work or rotating shiftwork also has the potential to lead to some adverse social effects. These effects depend strongly on individual cases, however, and should not be generalised. The following list provides an indication of what may occur:

- disruption of family life, such as meals and contact with spouse/partner or children. The ability of the family to adapt to shift work is an important factor in determining whether the worker will be able to cope with the work;
- participation in sport, clubs and groups is harder, and as the circle of friends is likely to be reduced, more individual forms of leisure may need to be pursued;
- lessening of social facilities (theatre, film, television, licensing hours), all of which operate for the convenience of those who work nine to five;
- employee relations problems may include "group feelings", communication of instructions from one shift to another and difficulty in obtaining general information on company activities; and
- the effects for women may be greater, due to their apparently greater deactivation rate than men, transport difficulties at night and the type of work they are frequently employed in, such as regulated assembly work. These issues do not imply that women are unsuited to night work, merely that employers should be aware of the different effects it may have on some individuals.

¶1104 Stand-by periods

As mentioned previously, in some occupations a large amount of time may be spent "on call". Often this is in addition to the worker's regular load, with some financial compensation for the inconvenience of being on stand-by.

Stand-by periods mean that an employee is not able to rest and fully recuperate from work. Such periods may add to the employee's stress load and create anxiety. Workers on stand-by have reported poor quality sleep, most likely due to the possibility of interruption. Where sleep is actually interrupted by a call, it is unlikely that the worker will recover the lost sleeping time.

Being on call may cause the worker's normal life (interaction with friends and family) to be disrupted or to take a lower priority than work, leading to possible isolation and a further source of stress. Where the worker is called away frequently, the social problems discussed above in relation to shiftwork will result.

Social Issues

To properly accommodate stand-by periods into a working schedule, it is necessary to consider them as a work-load. They are duty periods when the worker is not able to fully rest. Other periods should be allowed in order to ensure that the worker receives the necessary rest and recuperation to remain healthy in mind and body.

¶1105 Reducing the ill effects

The following list provides guidelines as to how night and shiftwork may be arranged in order to reduce the effects of some of the problems they cause.

1. Regular shifts may be preferable to rotating shifts, due to less fatigue and greater regularity of daily life. Where a rotating shift system is necessary, rapid rotation of morning, afternoon and night shifts with a maximum of three night shifts has produced significant improvements in health (demonstrated by both personal report and objective measurement) over workers engaging in the traditional three-weekly shift rotation system. (See ¶1102 for further details).

2. Workers should be given advice on how to cope with difficulties such as sleeping and meals (for example, how best to allocate "family time" and "sleeping time").

3. Shorter hours could be considered, either shorter shifts or preferably a shorter working week. This would better compensate for fatigue. Flexible working hours and/or additional annual leave could also be considered (note that a week's additional annual leave is a common award provision for shift workers).

4. Efforts should be made to reduce the number of people on night work to the essential minimum, through technology changes or attention to the distribution and operation of work tasks.

5. Retirement age could be lowered where superannuation schemes are provided. Susceptibility to fatigue increases with age, as does the period required to adapt to night work.

6. Improved facilities, such as transport, meal/cafeteria facilities and, where appropriate, beds.

7. Some people are attracted to night work initially because of the higher pay it offers, without considering whether they are really suited to it. Prior counselling may be a means of helping employees to adapt to the changes required by night work.

8. Employers should ensure that shift workers have access to employee services and facilities. For example, an answering machine could be provided so that workers can relay enquiries to the pay office/human resources department at night.

Tips on sleeping and eating for night workers

1. Plan sleeping times and do not change them. Make sure family and friends know these times.
2. If you have trouble sleeping enough in one stretch, try two: a longer sleep at the end of the shift, and a nap prior to returning to work.
3. Adjust your bedroom. Use heavy curtains, fan or air-conditioning, and no phone.
4. Do not eat a large meal just before sleeping. Also avoid alcohol, coffee and fatty foods that may prevent you from sleeping.
5. Do not substitute good meals with snacks and "junk foods".

Further references:

Shift Work, National Occupational Health and Safety Commission, 1991.

Guidelines for Workers and Management, Chamber of Mines and Energy of Western Australia, Inc, 1995.

Shiftwork — how to devise an effective roster, WorkCover NSW.

How to manage shiftwork, WorkCover NSW.

ALCOHOL AND DRUG DEPENDENCE

¶1106 Extent of the problem

Alcohol addiction is ranked as the fourth most prevalent disease in Australia after cancer, heart disease and mental illness. Alcoholics may be found at all levels of the work-force. The International Labour Office has reported that 15%-30% of all work fatalities are related to alcohol or other drugs, and employees with alcohol or drug problems have 200%-300% more absenteeism than other employees.

An alcoholic may be defined as a person whose physical and emotional health, job and general responsibilities towards society are repeatedly adversely affected by the continual abuse of alcoholic beverages.

"Occupational alcoholism" relates to recurrent poor job performance resulting from continued excessive drinking.

The problem appears to be most widespread in the age group 30-50.

Effects on work

Three main effects are characteristic of problem alcoholism amongst employees:

1. *Absenteeism*. This grows more pronounced as the problem deepens. Initially the problem drinker may try to avoid taking time off, but

later takes off more and more time including "off-the-job" absenteeism (prolonged lunch-hours, etc) and gives feeble excuses for absences.
2. *Accidents.* There is an increase in the accident rate both on and off the job.
3. *Job performance.* Initially job performance does not suffer it may even be above the average, as the employee tries very hard to do well in order to conceal the problem. However, if the alcohol dependence deepens, performance will fall off sharply.

It is harder to generalise about drug dependence because of the wide variety of drugs available. The effects and warning signs are likely to be similar to those attributed to alcohol, but note that there may also be the following additional problems:

- law enforcement agencies are involved in the problem to a greater extent than with alcohol; and
- while employees may begin work apparently acting normally, their behaviour may be unpredictable and present a high risk to both themselves and others.

Remember that the vast majority of alcoholics and drug abusers are in the work-force. It is also worth noting that dismissal of these employees without attempting to overcome their problems achieves nothing for industry. If the employee obtains another job, the problem has merely been recycled to another employer.

¶1107 Detecting the problem

There are various physical, psychological and social signs which, together with poor work performance, may help to identify someone with a drug/alcohol problem. Signs of problem drinking are set out in the chart below. The chart is reproduced with the permission of the former Australian Foundation on Alcoholism and Drug Dependence, from *Alcohol and Drug Dependence — A Joint Union Management Responsibility (1978).*

> **Some signs of alcohol problems at work**
> The whole emphasis of a successful rehabilitation program lies in the early discovery of clues pointing to an alcohol problem. This is when the alcohol-troubled worker will have the best chance of recovery. The following table not only deals with the nature of the signs of alcohol problems at work, but also compares the observations made by supervisors with the descriptions made by problem drinkers about their perceived clues.

Signs of developing alcoholism as reported by supervisors of alcoholics and alcoholics themselves		
Incidence	Observable (Supervisors)	Personal (Employee)
I Noticed early and frequently thereafter	Leaving post temporarily; absenteeism: half day or day; more unusual excuses for absences; lower quality of work; mood changes after lunch; red or bleary eyes	Hangover on job; increased nervousness/ jitteryness; hand tremors
II Noticed later but frequently thereafter	Less even, more spasmodic workpace; lower quantity of work; hangovers on job	Red or bleary eyes; more edgy/irritable; avoiding boss or associates
III Noticed fairly early but infrequently thereafter	Loud talking; drinking at lunch time; longer lunch periods; hand tremors	Morning drinking before work; drinking at lunch time; drinking during working hours; absenteeism: half day or day; more unusual excuses for absences; leaving post temporarily; leaving work early; late to work
IV Noticed late and infrequently thereafter	Drinking during working hours; avoiding boss or associates; flushed face; increase in real minor illnesses	Mood changes after lunch; longer lunch periods; breath purifiers; lower quality of work; lower quantity of work

(Trice, Christopher D Smithers Foundation, USA)

¶1107

Social Issues

> Note that Section I and to a lesser extent Section III are the most important ones and give early clues to problem drinking on the job.
>
> Other clues not mentioned in this table but that often come to the attention of the supervisory staff or medical personnel are domestic or financial difficulties.

In addition to those listed, other signs of a possible alcoholic include forgetfulness and memory blackouts, dishonesty, temper outbursts, intolerance and hypersensitivity about matters related to drinking, and suspicion of others, including management.

Of course, many of these signs can be symptomatic of non-alcohol related problems as well. Referral of an employee to an employment assistance program due to poor work performance can result in these problems being handled appropriately. The philosophy of such programs is to consider all problems that are affecting an employee's performance.

¶1108 Role of the employer

On a cost/benefit basis, various studies indicate that treating the problem is much more successful than dismissing and replacing the employee.

The most effective method of treating employees with alcohol and drug problems is to establish some sort of Employee Assistance Program (EAP). Some organisations use specific alcohol and drug programs, but EAPs have the advantage of taking a broad approach and dealing with any personal problems a worker may have. For this reason EAPs have become the preferred model of intervention into alcohol and drug problems at the workplace.

It is necessary to set up procedures for identifying and dealing with cases of alcoholic or drug abusing employees and, for reasons of consistency and fairness, ensure that these are observed.

The role of the employer is to identify the problem and encourage and arrange access to treatment. The actual provision of counselling services should be left to experts. An organisation may wish to establish internal counsellors, or it may use outside resources (see ¶1110). Part of the employer's role will be to ensure that supervisors have sufficient training to be able to identify employees in need of treatment. The chart at ¶1107 provides some guide to the signs supervisors should watch for.

¶1109 Employee Assistance Programs

Employee Assistance Programs (EAPs) are aimed at providing assistance to employees with any one of a wide range of personal problems, including drug and alcohol problems. Such programs are based on the premise that employees cannot and do not leave their problems at the front door of the workplace and then pick them up again on the way home. Personal problems of employees can affect their performance at work.

Personal problems that can be dealt with by an EAP include:

- drug and alcohol problems;
- marital and family problems;
- financial worries;
- gambling;
- interpersonal/social problems;
- health; and
- stress/lack of work satisfaction.

The key features of an EAP are:

- policy;
- procedures for referral; and
- counselling.

The employer may wish to publicise the availability of the program, and to establish a mechanism to evaluate its success.

EAP policy

The aim of any EAP policy is to set out the goals of the program, and to alleviate fears about the methods used. The confidential nature of any referral and counselling needs to be stressed.

The following features should be part of any EAP policy.

1. The aim of the EAP is to help employees restore their health and work performance to satisfactory levels.
2. The EAP is based on the philosophy that the employer is able to expect a satisfactory work performance, not that the worker's personal problems are the business of the employer.
3. No penalties of any sort will apply to any employee seeking assistance through an EAP.
4. Employees are encouraged to voluntarily refer themselves for assistance.
5. Supervisors should, as a matter of course, offer the EAP to any employee to whom they speak about poor work performance.

Social Issues

6. All details relating to counselling and treatment are to be treated with strict confidentiality.
7. Employees requiring treatment for a personal problem may use any accrued sick leave entitlements. Where no sick leave or annual leave is available, the employee will be entitled to leave without pay.

In addition, the policy may set out the range of problems dealt with by the EAP (see the list above).

Procedures for referral

Referral to an EAP is the key step in ensuring the success of the program. While supervisor-initiated referral may be the main mechanism of access, scope should always be made for self-referral to the EAP.

The following steps could be used by supervisors.

1. When interviewing an employee about work performance problems a supervisor should suggest that the employee is able to visit the EAP counsellor if there are personal problems that the employee does not wish to discuss with the supervisor. The decision whether to see the EAP counsellor remains with the employee.
2. If interviewing an employee for a second or third time, the supervisor should again offer the EAP to the employee, with a reminder that there could be disciplinary action if the work performance problems persist.
3. At the point of administering disciplinary action, the supervisor may either:
   (a) offer the EAP in lieu of disciplinary action; or
   (b) initiate disciplinary action and offer the EAP.

Some supervisors may feel reluctant to suggest the use of an EAP, as they feel the suggestion itself is a form of judgment. This can be overcome by stressing that the suggestion is based on a decline in work performance, and is in no way a moral judgment.

Where the employee elects to attend counselling during working hours, the supervisor is entitled to know that the employee attended the counselling (and no more). If an employee refers himself/herself to the EAP and elects to attend counselling outside of work hours, there is no need for the supervisor to be informed.

Further assistance

The Commonwealth funds a national tripartite organisation offering advice on EAPs and counselling to private workplaces. The service is offered under various names in the different States and Territories.

¶1110 Who is involved in counselling?

Counselling is at the heart of an EAP. Counsellors may be either specialist employees of the organisation, or belong to external agencies.

Counselling is aimed at helping individuals to identify strengths and problem areas in their lifestyles, and to make decisions about positive steps to resolve those problems.

EAP counselling is usually limited to one or two sessions. It is not designed to be long-term therapy. If further assistance is necessary, the counsellor may refer the employee to a specialist. Any further assistance will then be at the employee's expense.

Internal counsellors

Most commonly, counsellors are qualified psychologists with a clinical background or social workers with extensive counselling experience.

Counsellors need to have a background in a wide variety of personal problems, including drug and alcohol related problems. Other staff within the organisation, such as the human resources officer or medical staff, are not usually appropriate.

Internal counsellors must be perceived as having a high level of independence and credibility within the organisation.

Most organisations with fewer than 2,000 employees cannot justify the employment of a full-time EAP counsellor. In this case part-time counsellors may be employed, or the organisation may contract with an external agency.

External counsellors

When contracting with external counselling resources, it is usual for an organisation to pay an annual retainer rather than a fee for each individual. The latter option would compromise an employee's confidentiality if the employee chose to use the counsellor outside of work hours.

External counsellors may suffer from a lack of familiarity and contact with the organisation they are serving if sufficient orientation and access is not provided when establishing the relationship. *Further references:*

The Journal of Occupational Health and Safety — Australia and New Zealand, Vol 6, No 4, August 1990, p 265, "The development of alcohol and other drug programs in the workplace", CCH Australia Limited.

National Guidelines for Employee Assistance Program, National EAP Executive, Canberra, 1989.

Work, Drugs and Alcohol, Victorian Occupational Health and Safety Commission, 1992.

Drugs, Alcohol and the Workplace, WorkCover Authority of NSW

Drug Testing in the Workplace, Privacy Committee of New South Wales, 1993.

Not at Work, Mate (video and resource kit), the National Occupational Health and Safety Commission, 1994.

VIOLENCE AT WORK

¶1111 What is violence?

Violence is the unjust or unwarranted use of force and power. It can include verbal abuse, threats, harassment, physical assault, serious bodily injury and death. In its many forms, violence can occur in any occupation.

Workplace violence has been defined in various ways. Typical definitions include: "physical interpersonal aggression with the intent to harm or destroy" and "exertion of physical force so as to injure or abuse". Others define violence to include feelings of being threatened, abuse of managerial power, sexual harassment and gender issues. Common factors are that the violence occurs in conjunction with work, and that there is malicious intent and possibly physical injury. Other factors that could be included are psychological harm, and longer term outcomes of harassment and victimisation. The broader the definition, the more incidents will be captured.

A distinction has been drawn between workplace bullying and workplace violence. Bullying is said to relate to internal conflicts between employees, including managers and supervisors, and is located within an organisation. Workplace bullying manifests as persistent and repeated aggressive behaviours that give rise to feelings of victimisation in the person who is subjected to it.

In contrast, workplace violence is seen in a broad range of behaviours and circumstances that can be present both inside and outside the workplace. Unlike bullying, violence can be a single event. It can also be ongoing, random, persistent or premeditated behaviour. Violence can be from an internal or external source and an assailant or perpetrator may be known or unknown to the victim(s). Workplace violence has been used to describe abuse, threats, rape, assault or murder of employees by members of the public, disgruntled ex-employees or service users.

Those responsible for violence or bullying may be individuals, groups of individuals or the organisation itself may have an aggressive culture. Organisational assault could include placing employees in dangerous work situations or exposing them to the risk of emotional trauma as a result of unfair practices relating to retrenchment, downsizing or redeployment. It could also include allowing a climate of bullying or harassment to thrive in the workplace. An organisation may enable the use of email and the internet, where the medium is also used to send confronting images or abusive messages to individual employees or groups of employees. Some of these images and messages may also be considered as sexual harassment.

¶1112 Incidence of violence

Current research indicates the incidence of violence in the workplace. A *Bulletin/Morgan Poll* in May 1998 surveyed 641 people throughout Australia about their experience of aggression in the workplace. Fifty-two percent of men and 49% of women surveyed had experienced verbal or physical abuse from a member of the public while working, and 39% of men and 24% of women reported verbal or physical abuse from an employer or supervisor. A further 56% of men and 36% of women reported verbal or physical abuse from co-workers.

Some possible reasons for this level of violence in a variety of businesses and service industries could include:

- criminals and drug-affected people robbing alternative targets because cash-handling organisations, such as banks and building societies, have increased their security arrangements;
- people who are no longer eligible for government benefits and allowances as the welfare system is rationalised becoming increasingly desperate;
- domestic disputes spilling over to the workplace;
- psychiatric patients and juvenile offenders, being de-institutionalised and not provided with adequate support;
- disgruntled employees (past and present) of downsized organisations seeking revenge; and
- an increasing trend in society to accept the use of violence as a means to an end.

¶1113 Likely targets

A wide variety of people may be the targets of work-based violence either directly or indirectly. These include:

- people in public contact situations (nurses, counter enquiry staff, sales personnel, hospitality staff, and flight attendants);
- people working alone in community settings (real estate agents, security guards, taxi and bus drivers);
- staff handling cash or drugs (convenience stores, take away food outlets, video shops, petrol stations, pharmacies);

- those working within the legal system (judges, lawyers, child sexual assault workers, probation officers and police);
- private security staff (security guards and event management staff);
- those responsible for disciplining, relocating or terminating staff (HR personnel, security guards, supervisors and managers); and
- staff who are employed by organisations undergoing rapid and dramatic change (supervisors and managers).

A number of recent court cases reported in the media indicate an increasing incidence of violence inflicted against new members of organisations, such as apprentices and armed forces personnel, through de-humanising initiation and bastardisation practices.

¶1114 Risk management approach

An employer will need to exercise a duty of care in the context of workplace violence. There are four different circumstances that apply to workplace violence:

- protecting workers against harassment, aggression and violence from fellow employees, past and present;
- protecting workers from strangers, customers or estranged partners or family members;
- protecting the employer's customers/clients from violent acts by its employees; and
- protecting customers/clients from other people coming on to the premises.

To fulfil this duty of care, workplace policies addressing the issue of violence must address the three interlocking issues of violence prevention, management and post-trauma support. Policies and guidelines must include education and training, forward planning and a quick response capacity for crisis situations, together with on-going support for staff who experience violence in their workplace.

A risk management approach comprising the following elements is recommended.

Risk identification

This should be done with the cooperation of staff. A violence safety audit should be used to examine the following areas:

Policies and procedures — current codes of practice and safety policy; methods of induction; staff selection, training and supervision; monitoring of staff movements and current safety; staff support processes and availability; and incident recording and monitoring.

Location — working in isolated sites and working alone; disadvantaged neighbourhoods; difficult access to clients; reception areas; staff parking areas; interview rooms; and sites that may contain potential weapons.

Tasks — identification of staff pressure points; unacceptable task demands and rights and opportunities to abandon a task with unacceptably high risks.

Customers/clients — available information about client's previous history; matching experienced staff with aggressive clients; clients made aware of their rights and complaints procedures; and inappropriate client referrals.

Staff — appropriate aggression management training; negative staff attitudes to clients or other staff; backup resources available to staff; opportunities to withdraw from high-risk tasks and locations; and identification of staff targeted by stalkers or former partners.

After hours arrangements — arrangements for after hours backup; and level of assistance provided for getting to and from work safely.

Risk assessment

Once the areas of risk are identified, risk assessment must take place. This determines the factors that are contributing to the risk of violence, the risk severity in each situation, the urgency of preventative action and the priority for action that must be assigned to each risk.

Once the risk assessment is completed, management should develop or modify their program for dealing with aggressive and violent situations. This program should include six interrelated aspects:

1. organisation's operation;
2. training issues;
3. risk identification and assessment;
4. incident management;
5. post-incident management; and
6. ongoing monitoring and evaluation.

Risk control

Once the potentially violent situations have been identified and assessed, the next stage is to eliminate or control the risks by applying the control hierarchy of elimination, substitution, engineering controls, administrative controls and the use of personal protective equipment. Some suggestions follow.

Elimination — non-acceptance of dangerous clients or those affected by drugs or alcohol; close down the service or establish it elsewhere; outsource

¶1114

the service; screen out potentially violent employees; and implement a zero tolerance for violence policy.

Substitution — communicate by phone instead of direct face-to-face contact; select non-aggressive staff; use more experienced staff for difficult situations; and service non-aggressive clients.

Engineering controls — limit access; provide safe office/interviewing rooms; use electronic security devices; provide adequate external lighting; and provide safe and secure parking.

Administrative controls — develop clear policies and procedures; provide induction and ongoing training in aggression management; establish incident reporting; develop a crisis response plan; provide employee assistance programs; ensure staffing levels are adequate and rosters are realistic; and provide information about employers/employees rights and responsibilities.

Personal protective equipment — provide mobile phones and personal alarms where necessary.

By implementing the control strategies in cooperation with employees, violence in the workplace will be effectively managed, ensuring a safer and more secure environment for staff, clients and visitors.

Further information:

Coping with Violence — A Guide for the Human Services, Bowie, V. Whiting and Birch, London, 1996, Revised Second Edition.

Workplace Violence, Bowie, V. Australian Institute of Criminology

Conference Crime Against Business, Melbourne, June 1998.

Violence at Work, International Labour Office, Geneva, 1998.

Managing Violent and Potentially Violent Situations: A Guide for Workers and Organisations, Centre for Social Health Melbourne, Melbourne, Victoria, 1997.

Violence: A Risk Management Handbook for Dealing with Violence at Work. Grainger, C. Miintinta Press, Brisbane, Queensland, 1994.

Violence at Work. A Workplace Health and Safety Guide, Queensland Department of Training and Industrial Relations, Division of Workplace Health and Safety, 1995.

Minimum Standards for the Prevention and Management of Occupational Assault, Victorian Department of Human Services WorkCover Unit, 1996.

Safety and Health Solutions. Violence in the Workplace, WorkSafe Western Australia, 1996.

SMOKING AND THE WORKPLACE

¶1115 Passive smoking

Increased awareness of the health effects associated with passive smoking has made smoking at work a major issue. With so much time spent at the workplace exposure to tobacco smoke at work can have serious health consequences.

Smoking is now banned in most public sector workplaces, while smoking bans or restrictions are becoming increasingly common in the private sector. In 1990 the National Occupational Health and Safety Commission published the *National Policy Statement on Smoking and the Workplace*, that states that all Australian workplaces should aim to become smoke free. Note that smoking is now banned in many areas of public life (public transport, cinemas, domestic flights).

Tobacco smoke contains thousands of potentially harmful chemicals, including:

Cancer-causing agents (carcinogens). There are at least 43 known carcinogens in tobacco smoke. One of the most potent of all carcinogens is benzo-pyrene, discovered in tobacco smoke over 30 years ago.

Carbon monoxide. This is a poisonous gas that lowers the amount of oxygen carried by the blood.

Nicotine. This is the addictive drug that maintains the tobacco habit.

Radioactive compounds. The radioactive compounds found in highest concentration in cigarette smoke are polonium 210 and potassium 40. Radioactive compounds are known to cause cancer.

Hydrogen cyanide. In the amounts found in tobacco smoke this gas kills cilia, the tiny hairs that move together in waves to help keep our lungs clean.

Pesticides. A range of pesticides have been found in tobacco, including DDT, endrin, parathion and endosulfan.

Metals Many toxic metals, including arsenic and nickel, have been found in cigarette smoke.

Some tobacco smoke poisons are more concentrated in sidestream smoke (that smoke released directly into the air from the burning end of a cigarette) than in mainstream smoke (that smoke breathed in by the smoker). The concentration ratios are:

Social Issues

Poison	Ratio of concentrations Sidestream: Mainstream
Carbon monoxide	8:1
Nicotine	3:1
Formaldehyde (inhibits lung cilia)	51:1
Benzo-pyrene (carcinogen)	3:1
;gb-napthylamine (carcinogen)	39:1
4-Aminobiphenyl (carcinogen)	30:1
Dimethyl nitrosamine (carcinogen)	52:1

However, the concentrations of some of these substances in the air where people are smoking may be so low that the health hazard they present is negligible. The degree of risk depends on the degree of exposure.

Lung cancers and the increased risk of heart disease are the most serious consequences of passive smoking. The non-smoking wives of smoking husbands have been shown as having a significantly greater risk of developing lung cancer than non-smoking wives of non-smoking men in the same population. Some other consequences of involuntary exposure to tobacco smoke include the possibility of:

- aggravation of the condition of persons suffering angina pectoris, the pain caused by diminished blood supply to the heart muscles;
- precipitating attacks in asthma sufferers; and
- measurably impairing lung function in long-term passive smokers.

The links between passive smoking and adverse health effects are now well established. In February 1991 the Federal Court of Australia held that there was an established link between passive smoking and the development of lung cancer, asthma and respiratory diseases (*Australian Federation of Consumer Organisations Inc v The Tobacco Institute of Australia (1991)* ATPR ¶41-079). That case found that advertising published by the Tobacco Institute, to the effect that there was no scientific proof that passive smoking causes disease, was misleading and deceptive and therefore in breach of the Trade Practices Act. While the case did not concern health and safety at work, it can stand as a precedent for the link between passive smoking and the health effects mentioned above. On appeal, the Full Federal Court upheld the original judge's decision, and confirmed that the newspaper advertisement was misleading or deceptive.

¶1115

In the New South Wales District Court (27 May 1992), a jury found that passive smoking can be dangerous to health. It awarded $85,000 to a psychologist employed by the New South Wales Department of Health. She claimed that her employer had been negligent in exposing her to other people's cigarette smoke at her workplace. The smoke affected her pre-existing asthma and lung disease.

The Department of Occupational Health, Safety and Welfare (Western Australia) lost its prosecution against an employer (Burswood Resort Casino) on 17 September 1993. The Department alleged that the employer had breached the law by failing to control the level of tobacco smoke in the workplace, and exposing workers to damaging levels of passive smoking. The magistrate found that the prosecution had not proved that the employees' exposure to smoke was a health risk.

In the Australian Capital Territory, the *Smoke-Free Areas (Enclosed Public Places) Act* was passed in 1995. A code of practice for smoke-free workplaces also operates there.

Eye irritation (discomfort, itching, watering), nose and throat irritation, hoarseness, wheezing and headaches are commonly reported annoyances associated with the presence of tobacco smoke. People with allergies are especially susceptible to these irritations, and up to 20% of the population are included in this category.

There are certain workplaces where passive smoking may be particularly dangerous. Exposure to asbestos, radon, arsenic, aromatic amines and crystalline silica can interact with tobacco smoke to greatly increase the risk of lung cancer.

Other issues

Non-smokers' objections may be grounded in non-health issues, such as assertions regarding unpleasant odours in the workplace or in clothing, unsightliness and so on. Claims have also been made regarding the productivity of smokers — their absenteeism rates are higher and they take more breaks per day. It has also been asserted that smokers create additional costs in terms of cleaning and maintenance, property damage and depreciation.

Finally, there is the potential for friction between employees regarding the relative rights of smokers and non-smokers.

¶1115

¶1116 Some solutions

There is an obligation on the employer to provide a safe and healthy working environment but no specific legislation covering all workplaces in this area. The designing of an equitable solution, particularly where smoking may be a habit of many years for some employees, can present problems.

Many workplaces, in the absence of legislation, have developed workplace smoking policies. These may involve either banning smoking completely or restricting it to certain areas at the workplace. Either way, consultation with employees beforehand helps to ensure that the policy that is finally decided on is adhered to. The workplace health and safety committee may be an appropriate forum through which the consultation can be conducted.

Policies banning smoking

The most effective way to stop occupational health problems associated with smoking is to ban smoking from the workplace. However, as this is a fairly drastic step, it is likely to encounter some resistance if not implemented properly.

A smoking ban should be announced well in advance (at least three months) of the time it is to take effect. That time should be used to educate employees on the dangers of smoking (both passive and active) and to help them adjust. It may be possible to conduct an in-house quit program during that time, or to offer smokers paid leave to attend such a program.

The ban must extend to all levels of employees. A smoking ban that does not apply to executives would have little chance of acceptance.

Policies restricting smoking

Policies that restrict smoking to certain areas are common. Typically, smoking is restricted to enclosed private offices and a limited number of break areas. Air-conditioning systems may make this unworkable, as contaminated air is re-circulated to other areas. Common working areas are generally made smoke free. Smoking areas should be clearly marked, with the rest of the workplace being identified as smoke free.

Where a workplace is made up of open plan office space, it may not be possible for employees to smoke at their desks. The alternative is to group all smoking employees together, but this is not considered a very effective way of preventing passive smoking, and it may not be convenient.

As with policies banning smoking, reasonable notice in advance of the restrictions should be given. Education and quit programs may also be provided.

Further assistance

Cancer Councils, the National Heart Foundation and Health Departments in the various States may offer further assistance. Various private organisations offer smoking cessation courses designed for workplace implementation. The New South Wales Department of Health has produced suggested guidelines for a smoking at work policy in its publication *Towards a Non-Smoking Policy in the Workplace*.

Further references:

National Policy Statement on Smoking and the Workplace, National Occupational Health and Safety Commission, 1990.

Effects of Passive Smoking on Health, National Health and Medical Research Council, 1986

Australian Federation of Consumer Organisations Inc v The Tobacco Institute of Australia (1991) ATPR ¶41-079, Federal Court of Australia, 7 February 1991 (discusses the links between passive smoking and adverse health effects).

Tobacco Institute of Australia v Australian Federation of Consumer Organisations (1993) ATPR ¶41-199.

National Guidance Note on Passive Smoking in the Workplace, National Occupational Health and Safety Commission, 1994.

Going Smoke Free — a Guide for Workplaces, National Heart Foundation, 1991.

Passive Smoking in the Workplace, NSW WorkCover Authority, 1997.

Achieving Smoke Free Workplaces, Division of Workplace Health and Safety (Qld).

Smoke Free Workplaces Code of Practice (ACT).

¶1116

Social Issues

AGE AND EXPERIENCE OF WORKERS

¶1117 Issues relating to youth/inexperience or age of workers

One of the central principles of the modern approach to occupational health and safety is that the job should be made safe for the typical range of workers, rather than restricting personnel to only a select number who can cope with the working conditions. Within this context employers need to take into account the various characteristics of workers, including their age and experience, to ensure a safe and healthy working environment and systems of work that are safe for all workers.

In recent years there has been considerable publicity about workplace risks to younger workers. National statistics indicate that more than 30 young people (under the age of 25) are killed at work each year in Australia. Thousands are injured, and a proportion of these are left with a permanent disability of some kind as a result. For example, in New South Wales alone, 858 young workers suffered permanent disability from an employment injury in the 1995/96 financial year.

The rate of injury to young men is particularly high — approximately three times higher than for young women. Data from the National Occupational Health and Safety Commission indicates that young male workers, particularly apprentices, have a one in two chance of being injured at work each year, compared to a one in eighteen chance of injury for older workers.

Accident rates for young people driving vehicles are known to be higher than rates for older people — hence the higher cost of car insurance if the vehicle is to be driven by a person under 25. This higher risk is reflected in the work context as well. Young men in particular can be over-confident in their ability to, for example, drive a forklift truck even though they may not have any experience with it and have not been trained in its safe operation.

Young workers (at their stage in life) are particularly at risk. They are unfamiliar with the workplace and its hazards. They can be influenced by peer pressure. They may feel, for example, a lack of interest or respect for the idea of "safety first", have a tendency to horseplay, or bravado that includes taking unnecessary risks.

Youth and inexperience in the workplace can also make young workers more prone to other workplace dangers such as bullying, victimisation or other oppressive, distressing or dangerous practices such as "initiation rituals". Employers must take factors such as these into account when defining safe systems of work. Training and supervision are particularly important in the case of young and inexperienced workers.

With older workers the issues are different. Some employers are concerned that older workers may be less responsive, more difficult to train, have higher accident rates and take more time off work due to age-related illnesses. These invalid assumptions have significantly affected the job opportunities for older Australians at a time when there is a significant shortage of skilled workers.

Managing older workers, at least from an occupational health and safety viewpoint, should be no different than managing any other group of workers. In fact, as measured by many outcomes, older workers can be a very cost-effective resource for employers because of:

- accumulated skills;
- lower training costs;
- higher retention levels;
- good work ethics;
- excellent customer service skills;
- low levels of sickness or absence due to personal or family member ill-health;
- low likelihood of workers compensation claims;
- significantly reduced absenteeism due to marital upsets; and
- lower absenteeism due to sport accidents and motor vehicle accidents.

¶1118 Employer concerns

The most obvious aspect of young people in terms of their risk of workplace accidents is their physical immaturity. Developing bodies are more vulnerable in some ways to damage from manual handling of heavy loads, fatigue or other stresses.

Other characteristics of youth can also contribute to increased risk — consider the following:

- Unless taught otherwise, they tend to assume the risks are part of the job.

- Some working environments can be alien and intimidating to those who are not familiar with them.

- They may not recognise the potential for hazards in the working environment, due to their limited experience with it.

- They may be inhibited about asking questions, saying they do not understand instructions, or objecting if they think something looks dangerous.

- They may fear looking ridiculous or weak in front of their peers, and may resist wearing personal protective equipment (PPE) — especially if older co-workers neglect to wear PPE when appropriate.

- They have a greater tendency to "skylark" or take unnecessary risks.

- They may be experiencing a lifestyle change, because they have entered the workforce. This could mean more exposure to drug and alcohol use and more ready cash.

- They want to fit in, and do not always recognise their own limitations.

- They tend not to have proper sleep patterns and forego nourishing meals due to social pressures.

Not all workers under 25 fit into the same mould, but employers need to consider the risks to which young people are exposed in the light of the above-listed tendencies. Another issue to take into account is the frequent tendency for young people to work while also undertaking a course of study — going straight from school or college to long shifts at work increases their risk of fatigue.

Because young people are often new to the workplace, employed on a casual or short-term basis and not familiar with all of the rules and procedures, they can be unaware of safety issues and of the employer's attempts to address health and safety. For example, they can be under-represented on workplace health and safety committees, and not oriented to the need to report risks. This again underscores the need for sufficient training and supervision of young people in the workplace.

In relation to older workers, employers may be concerned about physiological changes. Ageing by itself is not a disease, but some of the

¶1118

inescapable age-related physiological changes are associated with an increased prevalence of pathology. Age-related declines include:

- reduction in hearing acuity
- reduced visual acuity (even after adjustment with eyeglasses) as well as differing sensitivities to coloured light such as the multicoloured displays on some VDT applications
- reduced maximal aerobic capacity and an increase in the rest time needed after performing physically demanding work.
- reduced muscle strength, including hand grip
- reduced speed of processing new information
- decreased tolerance to heat or cold in the working environment
- increased prevalence of degenerative musculoskeletal changes including arthritis and loss of bone mass
- increased prevalence of cardiovascular problems, both due to underlying disease such as hypertension, as well as the possible effects of medications
- increased risks of undetected metabolic diseases such as diabetes
- increased prevalence of dementia
- increased difficulty in coping with shift work and extended working hours.

Among this array of potential age-related disorders, a large degree of individual variation will occur. The healthy worker effect is especially prominent in older workers, with clear evidence of self-selection. Fitter, older workers often choose to work to an older age. Those with serious underlying health problems do not. A physically active, healthy 70-year-old man may have much greater physical strength than a 45-year-old woman.

These individual differences increase with age, and clearly require fitting the individual to the job. This is essential for workers aged over 50. In addition to assessing the physical, mental and social requirements of the task, a responsible employer will focus on the individual's work capacity or functional age, rather than just rely on the very crude indicator, chronological age.

Surveys undertaken in Europe, the US and in Australia have consistently shown that employers recognise the value of older workers, and acknowledge their lower absenteeism, greater work ethic, thoroughness and

Social Issues

reliability, etc but consistently (and erroneously) believe that older workers have more accidents and cannot be retrained for new tasks.

Accident rates among older workers are higher in certain classes of employment. Older workers have higher accident rates due to slips and falls and are involved in a higher percentage of transportation-related job fatalities. However, in the vast majority of post-industrial Australian workplaces, these potential problems can be avoided.

With proper floor design, non-slip surfaces, adequate guard rails and appropriate use of safety harnesses, the working environment can be made safer for all workers, not just those younger or older than some arbitrary figure. The transportation fatality rates, especially in Europe and the US, are largely caused by the much higher percentage of older owner-drivers driving their own semi-trailers and operating as independent contractors. Lifestyle issues contribute to a higher accident rate in this group of workers. These issues include high levels of cigarette smoking and hypertension, relatively low driving skill levels, older mechanical equipment, excessive tiredness and the financial necessity to keep working to demanding schedules to earn a decent return.

Health promotion and screening programs, coupled to rigorous enforcement of safe working hours, use of log books and the recent introduction of satellite tracking of heavy transport vehicles should all lead to lower accident rates in this group of workers.

Concerns about the training of older workers have a long history. However, it has been demonstrated that older people can learn as fast as younger ones if appropriate teaching methods are adopted.

An American report examined the cost/benefit of retraining older workers versus replacing them with younger workers. The analysis showed that the cost of retraining 20 older engineers (half of them in their 50s) was only one third of the cost of replacing them with younger employees. Another study concluded that forcibly retiring workers aged over 55 was over 30% more expensive than retraining and continuing to employ the older workers.

These studies and others were summarised by the US National Foundation for Occupational Health Research in 1990. The foundation concluded that there is no clear-cut evidence that older workers show a significant decline in their skills or capacities, and in certain areas they can match or outperform their younger counterparts.

¶1118

¶1119 Managing risks for young/inexperienced and older workers

The provision of adequate training and supervision is the foundation of managing the health and safety of young or inexperienced workers.

Induction training as soon as a young person joins an organisation — before they commence duties — should build on any vocational training the employee has had at school or college and should introduce the worker to the health and safety risks inherent in the job.

It is important to focus on the job the worker will be performing, rather than just the employer's premises. For example, young bicycle couriers have been hit by cars, or young people injured in fast-food home delivery. Though the injuries did not happen on the employers' premises, risks on the road are still part of the full set of risks to which the worker is exposed when carrying out the employer's business, and so should receive just as much attention during induction training. All workers, and especially young workers, should receive induction training that covers the following matters:

- the organisation's health and safety policy;
- hazards present in the workplace, or involved in work procedures;
- relevant safe work practices;
- behaviour likely to lead to danger or injury;
- any documentation the worker should be aware of (such as operating manuals and material safety data sheets);
- maintenance and service routines (for example, when to request maintenance and who to ask);
- any special safety information needed (such as precautions for working under certain conditions, how to use emergency stop buttons or other safety devices, and rules about use of drugs and alcohol);
- the types of personal protective equipment needed, together with information on fitting, using, maintaining and storing this equipment;
- emergency procedures;
- reporting requirements for injuries or incidents; and
- where relevant, details on how accidents have occurred in the past involving the same work process.

On-the-job instruction in the particular tasks to be performed is an essential follow-up to formal or general training. This should recognise that young persons cannot be expected to be aware of all the dangers that more experienced or mature workers might know and avoid. The question of competence to drive or operate machinery is especially important. Young men in particular need to understand that they are not free to use a vehicle if they have not been trained in its safe use — even if the key has been unwisely left in the ignition.

Supervision of all employees, especially the young, in the first few weeks of work, is vital to ensure that training and instruction has been understood and safe working practices are being adhered to. Young workers may adopt role models among older or more experienced workers. The example presented by other workers and peer pressure to emulate other workers' practices makes it clear that the full context of training has to be considered. Formal training can easily be undermined if it is contradicted by the "informal training" that happens when the young worker observes how the other workers do things.

The importance of adequate instruction and supervision for young workers highlights the need for adequate supervisor training. Supervisors must be competent not only in the quality of initial instructions to workers, but also in terms of their follow-up observations of how young workers perform the tasks. They also need to recognise the effects of peer pressure and general work practices. Supervisors must be able to achieve the same level of compliance with safe work practices among older workers as they expect from younger workers.

Managers in turn need to assess the adequacy and effectiveness of supervisor training, as well as the suitability of induction and on-the-job training. They also need to monitor compliance with safe operating procedures in which workers have been trained.

Managing older workers should be little different from managing any other group traditionally considered to have special needs, such as women, persons of non-English speaking background, people with disabilities and different cultural groups eg. indigenous Australians. Being a good people manager implies an ability to ensure that a company's needs are effectively married to its workers' needs to ensure optimum outcomes for all parties. In practice, however, managers need to be aware of some of the sensible and cost-effective workplace adjustments that should be made to accommodate older employers.

¶1119

Vision screening

At the simplest level, employers with visually demanding tasks should consider the implementation of visual acuity screening tests for employment applicants. Since eyesight gradually deteriorates with age, a significant percentage of older workers will need to have their vision corrected to perform their jobs safely and efficiently.

While it may be obvious to everyone that an older driver of a heavy transport vehicle, eg a petrol tanker, will need excellent (corrected) eyesight, many other employees in the jobs working with Visual Display Terminals (VDT) might also need sight correction. Frequently, vision testing only focuses on reading distance and infinity. However, VDT workers need to be able to focus clearly at the VDT viewing distance which may be 500mm to 800mm from the eye. Traditional refractions will not compensate for this viewing distance, and employees with prescription eyeglasses need to advise their optometrist or ophthalmologist of the focal distances that they will be working at.

Employers of large workforces in the VDT industry in the US and Europe provide free eyesight screening, and either free or greatly subsidised prescription eyeglasses to help their employees see most clearly at the required VDT viewing distance. These in-house programs are of benefit to all workers, but especially older workers whose uncorrected eyesight accommodation will have deteriorated with age. This also assists the employer in reducing the number of errors resulting from poor recognition of work.

Health promotion programs

Health promotion programs can substantially benefit all employees, and especially older workers. Healthy older workers will greatly benefit from periodic screening programs covering areas such as glaucoma detection, diabetes screening, hypertension screening and screening for certain types of cancers, especially those in the breast, prostate and colon. These programs have widespread community acceptance and may also significantly benefit workplace morale and the corporate image. The programs have been shown internationally to be cost effective. They can reduce absenteeism by assisting workers to attend to their health.

If a company has a long term and ongoing investment in health promotion at the workplace, a substantial amount of valuable health data can be collected. This enables early detection of departures from normal with consequent improvements in treatment and health outcomes.

¶1119

Ultimately, this will improve worker productivity through lower absenteeism, and greater feelings of wellbeing.

One controversial area for employers to consider is the issue of influenza vaccination for workers. National Health and Medical Research Council data basically suggest that even healthy older workers will benefit from yearly influenza vaccination. The problems are not so much just being sick with influenza for a few days, but the serious health consequences such as pneumonia that can result from a bout of influenza. While this is of major concern to older workers, many younger workers are also at risk, especially if they have pre-existing chronic diseases or some immune system compromise. This is a good example of where designing a program for older workers can have significant benefits for all workers.

Shiftwork

Many workers liken the effects of shiftwork to chronic jet lag. Numerous studies have shown that older workers do not tolerate forced shiftwork as well as younger workers. There are many reasons for this, including difficulty in sleeping, adjusting to altered day/night cycles, gastrointestinal upsets and circadian dysrhythmia affecting conditions such as diabetes mellitus and hypertension, etc.

Many older workers select themselves out of consideration for shiftwork jobs. Their financial needs are usually lower, and the extra money from shift loadings, etc, is just not worth the side effects of shiftwork. It is suggested that wherever possible, shiftwork requirements be met entirely with volunteers and that older workers be given reasonable consideration in relation to requests for transfer to permanent day work. Shift rotation and length should be structured to reduce the impact on individuals.

Training and retraining

Old dogs can learn new tricks, and the principles of adult learning have been understood for nearly a century. While our technology is changing rapidly, older workers in particular may feel marginalised when their 20-plus years of experience suddenly becomes valueless. The occupation of typesetting is an example. This was a trade that had been fairly stable for 500 years, yet in less than one generation, the technology requirements of today's publishing have meant that the manual setting of type has become a lost art.

What is needed in retraining workers, particularly older workers, is the opportunity to experiment, to make mistakes, and learn by doing. In the case of computer technology, personal computers have been around for less than 20 years, and the World Wide Web for only 5 years. Any mature professional, including some aged in their 30s, may therefore not have been exposed to these technologies at school, TAFE or university.

Quality training at an appropriate pace, frequent reinforcement and support, and the ability to experiment at home or at work in a low risk environment will help older workers to become proficient in the use of the newer technologies. Once acquired, these new skills will enable older workers to successfully meet the technology challenges of today's business environment.

Further information:

Managing an Ageing Workforce, Patrickson M and Hartmann L (eds). Australia, BPP Publishing, 1998.

Are young workers a greater risk? Factsheet number 2, Workplace Health and Safety Division, Queensland Chamber of Commerce and Industry, 1999

Young workers, work experience and work placement students — don't let their first day be their worst day, WorkCover NSW, 1997.

¶1119

Chapter 12

Reference Section

Introduction	¶1201
Government organisations	¶1202
Other organisations	¶1203
Publications	¶1204
Audio-visual aids	¶1205
Training and tertiary courses	¶1206

¶1201 Introduction

Throughout this book, it has been emphasised that occupational health and safety covers many areas and involves a large number of technical issues. Multi-disciplinary knowledge is required to cover all aspects of the subject. The aim of this book is not to cover complex issues in full detail, but to act as a starting point for practitioners to seek further details for themselves.

Every organisation will face different problems and circumstances at its own workplace and will therefore have differing needs for information in order to resolve problems. Different solutions may also be appropriate.

In recognition of all these factors, the purpose of this chapter is to provide a guide to the sources of both general and specialised assistance available to organisations. The chapter describes the range of government and private organisations in the health and safety field, and some general publications, audio- visual aids and courses.

¶1202 Government organisations

Commonwealth

National Occupational Health and Safety Commission (formerly known as Worksafe Australia)

The Commission was established as a statutory body in December 1985. It comes within the portfolio of the Minister for Workplace Relations and Small Business.

The Commission is based in Sydney. It is, in effect, a tripartite committee (the three dimensions being State/Federal Government, employer and union, as detailed in the membership listing below).

The Commission has 18 members: a Chairman, a Chief Executive Officer, three ACTU and three ACCI representatives, nominees of each State and Territory and two Commonwealth Government nominees. It provides a forum for developing the national strategy and determining national priorities in implementation of national policy. Members of the Commission hold office for three years.

Under the *National Occupational Health and Safety Commission Act 1985*, it is charged with a range of specific functions. These include formulation of policies and strategies, encouragement of co-operation between Commonwealth, State and Territory Governments and groups and individuals; declaration of national standards and codes of practice; collection, interpretation and dissemination of information; conduct of inquiries; provision and assistance with training in occupational health and safety matters; publication of reports and papers; provision of grants; support of research and testing, and encouragement of the application or use of the results.

The Commission also has responsibility for the administration of the *National Industrial Chemicals Notification and Assessment Scheme*. This requires the assessment of new industrial chemicals prior to their use in Australia, and for reviews of certain existing chemicals.

Since its inception the National Occupational Health and Safety Commission has developed into the primary occupational health and safety body in Australia. It has developed national standards and codes of practice that have considerable weight as guides to acceptable practice. The Commission does not, however, have the power to legislate in this area.

The National Occupational Health and Safety Commission publications (see ¶1204) are available through Commonwealth Government Bookshops. Visit their website at http://www.nohsc.gov.au.

Reference Section

Comcare Australia

Comcare Australia is the workers compensation insurer for the federal government, providing health and safety, rehabilitation and compensation services to federal employees. It has its head office in Canberra and offices in the other States and Territories. It administers the *Occupational Health and Safety (Federal Employment) Act 1991*, and is answerable to the Federal Minister for Workplace Relations and Small Business.

Comcare does not have its own inspectorate, but manages its inspection functions through cooperative arrangements with each of the States and Territories. Visit their website at http://www.comcare.gov.au

Department of Community Services and Health

This Department has responsibility for several occupational health matters. The *Australian Radiation Laboratory*, the *National Acoustic Laboratories*, the *Commonwealth Rehabilitation Service* and the *National Health and Medical Research Council* come within its administration.

Australian Nuclear Science and Technology Organisation (ANSTO)

ANSTO is Australia's national nuclear technology and research organisation. The organisation is located at Lucas Heights, on the outskirts of Sydney, where it operates two research reactors. The organisation is heavily involved in occupational health and safety, providing expertise in all areas of nuclear safety and conducting a wide range of general safety courses.

Other Government organisations

Additional Commonwealth Government Departments or bodies with some involvement in the field include the *Commonwealth Scientific and Industrial Research Organisation* and the *Commonwealth Fire Board*.

State Government Departments

Most of the relevant legislation is State Government sourced. The principal authority responsible in each State and Territory is given below:

NSW:

WorkCover Authority of NSW

Website address: http://www.workcover.nsw.gov.au

Vic:

Victorian WorkCover Authority

¶1202

Website address: http://www.workcover.vic.gov.au

Qld:

Workplace Health and Safety, Department of Training and Industrial Relations

Website address: http://www.detir.qld.gov.au/hs/hs.htm

SA:

WorkCover Corporation of South Australia

Website address: http://www.workcover.com

WA:

WorkSafe Western Australia

Website address: http://wwwl.safetyline.wa.gov.au

Tas:

Workplace Standards Authority

Website address: http://www.wsa.tas.gov.au

ACT:

ACT WorkCover

NT:

Work Health Authority

Website address: http://www.nt.gov.au/wha

Other departments may administer legislation on subjects such as mining, water, noise and air pollution, disposal of toxic wastes, approval for electrical equipment, agriculture and poisons. Note that some matters may fall within the administrative control of local government councils.

¶1203 Other organisations

The list of organisations here is not meant to be exhaustive. In addition to the ones listed here, there are specialist organisations and a wide range of consultants who advise on particular topics such as lighting, acoustics, stress management techniques, harmful substances, air-conditioning, etc. There are also a number of organisations that provide useful software.

¶1203

Reference Section

Standards Association of Australia (Standards Australia)

The Standards Association of Australia is a non-government, independent body incorporated by Royal Charter. It is known by its trading name of Standards Australia, and is referred to as such throughout this book.

The Association's main function of investigation, development and recommendation of standards includes standards relevant to occupational health and safety. These standards are prepared by a process of voluntary consensus and many have been called up in State legislation. When Australian Standards are called up in legislation they become laws. In other circumstances, standards are regarded as authoritative or, at least, as excellent guidelines to the appropriate quality or procedure.

Responsibility for the development of national standards in occupational health and safety lies with the National Occupational Health and Safety Commission. It has established a Standards Development Standing Committee whose task includes the development and review of standards that may or may not incorporate standards produced by Standards Australia.

Copies of all Australian Standards are available over the phone from the Association's National Sales Centre, directly from the State offices of the Association or from their website, http://www.standards.com.au. Charges are based on the size of the particular standard.

The Association's publications include a *Register of Australian Standards* referenced in State and Commonwealth legislation, a listing of codes and Standards that affect occupational safety, a monthly magazine, *The Australian Standard*, and a publication on protective equipment suitable for TAFE and safety course students, the *Manual of Industrial Personal Protection*.

National Safety Council of Australia (NSCA)

The National Safety Council of Australia (NSCA) is a single incorporated national not-for-profit organisation, funded by a combination of members' subscriptions and discounted fees for services provided to members and non-members.

NSCA has branch offices throughout Tasmania, New South Wales, Victoria and Queensland, with formally affiliated offices in South Australia, Western Australia and the Northern Territory. It offers professional consultancy services and guidance on the initiation and implementation of occupational health and safety management programs. It is the largest organisation of its type within Australasia, and its main aim is the continuous improvement of occupational health and safety in both the private and public sector.

¶1203

Services offered include health and safety surveys, audits, scientific services, quantified risk assessments, a wide range of management and employee training, risk management programs and behavioural change processes. NSCA also offers several awards as incentives to organisations and individuals. The winners of these awards are published in NSCA's national magazines *Australian Safety News* and *Home Safety and Security*, thereby recognising their achievements in the reduction of the risk of work-related injury and illness.

Safety Institute of Australia (SIA)

The Safety Institute of Australia Inc is a voluntary organisation with a membership drawn from people carrying out occupational safety and health functions in a wide range of industrial and Government employment. The Institute aims to promote the science and practice of accident prevention, to facilitate the exchange of information and to encourage the development of safety as a profession. Besides a Federal Council that co-ordinates activities, there are divisions in each State and the ACT.

Australasian Faculty of Occupational Medicine (AFOM)

The Faculty, inaugurated by the Governor-General in 1984, now has members worldwide (mostly from Australia, New Zealand, Singapore and Hong Kong). It is concerned with developing high standards of occupational health practice and runs intensive training programs enabling medical doctors to proceed from trainee to fellowship status.

Ergonomics Society of Australia

The Ergonomics Society of Australia is concerned with the interrelationship of human beings with their work tasks, their working environments and the equipment they use. As this study encompasses medical, psychological, scientific and engineering disciplines, this spread of interests is reflected in the Society's membership. The Society holds branch meetings and national conferences.

Australian Institute of Occupational Hygienists (AIOH)

The Australian Institute of Occupational Hygienists aims to promote its profession, to improve the knowledge and practice of its members and to represent them at national and international levels. The Institute maintains a register of professional occupational hygienists.

¶1203

Australian and New Zealand Society of Occupational Medicine (ANZSOM)

The Australian and New Zealand Society of Occupational Medicine is generally open to medical practitioners involved in occupational medicine. The Society seeks to educate people in the proper practices of occupational health.

Australian Occupational Health Nurses Association (AOHNA)

The Association aims to promote and maintain high standards of occupational health practice in Australia. This is achieved through education, policy advice and development, research, exchange of ideas and conferences and meetings.

Australian Fire Protection Association

The Australian Fire Protection Association Ltd is a non-profit, technical and educational organisation with the aim of safeguarding life and property against fire. The Association maintains a technical and educational reference library and a film library of instructional films that are loaned free to members. Training courses and seminars are conducted, and bulletins and a journal are produced.

St John Ambulance Association and Australian Red Cross Society

Both organisations conduct certificate courses in first aid, that are acceptable as an approved level course under safety legislation. Courses can be arranged within an organisation (if numbers are sufficient) or for a combined group.

Association of Risk and Insurance Managers of Australia (ARIMA)

The objectives of ARIMA are to promote the profession of risk management and insurance in Australia through education and exchange of ideas. The Association has close links with the Risk and Insurance Management Society (USA) and the Association of Insurance and Risk Managers in Industry and Commerce (UK) and is a member of the International Federation of Risk and Insurance Management Associations.

Australian Council of Trade Unions (ACTU)

The ACTU has adopted an official policy on occupational health and safety, that covers ACTU involvement in legislative change, research, education and training, managerial action and direct union activity. It has also established an Occupational Health Unit. The Unit publishes a regular newsletter, as well as guidelines on particular subjects, such as VDUs and manual handling.

¶1203

The ACTU has three representatives in the National Occupational Health and Safety Commission.

State Labor Councils and Unions

State labor councils and a number of unions have OHS/Welfare departments that provide information on OHS issues relating to their members. This information is often distributed through newsletters or publications. At times these bodies will also undertake or fund studies into particular OHS concerns of their members.

Trade Union Training Authority (TUTA)

TUTA is the registered training provider for the ACTU Organising Centre. It conducts courses for trade union officials, delegates and members, several of which deal with occupational health and safety either solely or in part. Short courses (about three days) are regularly available in capital cities and regional centres.

Workers' Health Centres

Workers' Health Centres generally specialise in the provision of occupational health and safety services to union members. The centres are also involved in the promotion of safer workplaces. Centres are located in Sydney, Wollongong, Newcastle, Melbourne and Brisbane.

Australian Chamber of Commerce & Industry (ACCI)

The Australian Chamber of Commerce & Industry is the largest single organisation representing industry and commerce at a federal level. The ACCI is the employer representative on the National Occupational Health and Safety Commission (NOHSC).

The ACCI Occupational Health and Safety Facility provides information on occupational health and safety to employer organisations with literature and database searches. They develop and distribute information pamphlets on occupational health and safety issues, and distribute NOHSC draft standards and codes of practice for information and comment by employers.

Employee Assistance Programs

Employee health and welfare assistance programs are available from organisations for private workplaces. The services available include consulting, training and counselling on problems in the workplace including drug and alcohol and work-related post-accident trauma. For further information, contact the occupational health and safety or industrial

¶1203

relations authority in your State or Territory. See ¶1109 for more information on EAPs.

¶1204 Publications

It is not possible to provide an exhaustive list of the available publications. The following is a guide to a range of useful general references, books, and periodicals.

References

Australian Occupational Health & Safety Law (3 vols), CCH Australia Limited, Sydney

This loose-leaf reporting service emphasises the legislative side of occupational safety and health. Legislation is reproduced either in full text or as extracts. It contains commentary on legal liability, planning, promotion and rehabilitation, administration and enforcement and particular health problems. It also includes a comprehensive reference and advisory services section.

Managing Occupational Health & Safety, CCH Australia Limited, Sydney

This publication is a loose-leaf service that aims to provide a guide for conducting and administering occupational health and safety at the workplace. The manual does not set out legislation, but aims to provide a practical approach to dealing effectively with occupational health and safety. The manual's sections include a guide to the human body, hazard control, training, first aid, emergency planning and record keeping, as well as dealing with a variety of specific occupational safety problems (hearing conservation, electrical safety etc)

National Occupational Health and Safety Commission publications

The National Occupational Health and Safety Commission publishes a range of national standards and codes of practice dealing with specific health and safety issues. In addition there are information brochures, research reports, resource kits and workplace guides. The Commission has taken over the publication of the "Working Environment" series, originally produced by the (then) Department of Employment and Industrial Relations. This series includes titles such as *Amenities at Work, Clean Air at Work, Fire Safety at Work, VDUs at Work*, etc.

The National Occupational Health and Safety Commission publications are available from Commonwealth Government Bookshops or in libraries with special OHS collections.

Occupational Health and Safety in Australia — A Guide to Sources of Information, Margaret Casson, Techpress, Stepney, SA, 1993 (3rd edition)

As its name suggests, the book aims to cover the relevant information sources in the fields of occupational safety and health. It includes information on literature, organisations, journals, abstracting services, bibliographies, research, reports, conferences, yearbooks, manuals, dictionaries and encyclopaedias, audio-visual materials, statistics and legislation.

National Occupational Health and Safety Directory, Newsletter Information Services, Sydney

This publication, with a new edition available every year, provides a guide to all those operating in the field of occupational health and safety. Names, contact details and descriptions of services are given for Government bodies, consultants, insurers, health professionals, training providers and equipment providers.

Encyclopaedia of Occupational Health and Safety, ILO, Geneva

The fourth edition of this comprehensive reference work was published in 1998. Written by 900 specialists, it covers a huge range of occupational health and safety topics. Copies are held by the Commonwealth and State Government Departments concerned, as well as by many voluntary organisations. For inquiries or availability, contact Hunter Publications, Melbourne.

Periodicals

A comprehensive list of international periodicals appears in the reference section of CCH's Australian Occupational Health & Safety Law (see above). The following is a list of useful Australian journals.

1. *The Journal of Occupational Health and Safety — Australia and New Zealand*, CCH Australia Limited.
2. *Australian Safety News*, National Safety Council of Australia (some State branches of the Council also publish their own newsletters).
3. *OH&S: the Australian Journal of Workplace Health and Safety*, Scriptographic Publications, Sydney.
4. *Occupational Health Magazine* and *Occupational Health Newsletter*, Newsletter Information Services, Sydney.
5. *Ergonomics Australia*, Ergonomics Society of Australia, Qld.
6. *Fire Journal*, Australian Fire Protection Association.

¶1204

7. *The Australian Standard*, Standards Australia, Sydney.
8. *WorkCover News*, NSW WorkCover Authority, Sydney.
9. *Workwords*, Victorian WorkCover Authority, Melbourne.
10. *Health and Safety Bulletin*, WorkCover Corporation of South Australia, Adelaide.

Books

CCH's *Australian Occupational Health & Safety Law* (see above) contains a comprehensive book list, divided into subject titles. *The Journal of Occupational Health and Safety — Australia and New Zealand* (see above) features book reviews including summaries of Government reports.

A selection of useful general titles is provided below.

Workers Compensation Law and Practice in New South Wales, 3rd edition, F Marks and B McLean, CCH Australia Limited, Sydney, 1992.

This book covers administration, legislation and case law developments in New South Wales.

Work and Health, M Quinlan (ed), MacMillan Education Australia, Melbourne, 1993.

Discusses the origins, management and regulation of occupational illness.

Evaluation of Human Work: A Practical Ergonomics Methodology (edited by Wilson and Corlett), Taylor and Francis, London, 1995.

An ergonomics text aimed specifically at applying ergonomics in the workplace. It comprises 34 chapters written by 38 contributors.

Managing Occupational Health and Safety in Australia, Quinlan and Boyle, MacMillan, Melbourne, 1991.

A text for those who already have a broad understanding of health and safety issues. An analysis of the state of occupational health and safety in Australia is presented.

Occupational Medicine (edited by LaDou), Appleton and Lange, US, 1990 (Australian distributor: McGraw Hill).

A book aimed at providing general practitioners with an introduction to occupational medicine. Of some interest also to practising occupational physicians.

¶1204

¶1205 Audio-visual aids

Videos covering occupational health and safety topics can provide very powerful communications. Interest in the area has seen a rapid expansion of releases produced independently and in association with various interested organisations. *The Journal of Occupational Health and Safety — Australia and New Zealand* (see ¶1204) includes video reviews with most issues.

The National Occupational Health and Safety Commission has a comprehensive collection of OHS videos and publishes a catalogue of these.

Many of the other organisations listed in this chapter will be able to provide videos on general or particular subjects, or else be able to advise where such items may be obtained. In addition, there are private organisations that rent and sell safety videos.

¶1206 Training and tertiary courses

Training courses can be divided into three groups, Government-run courses, those conducted by private organisations, and courses run at educational institutions. As with other areas of occupational health and safety, there has been a rapid expansion in training and courses for all levels of interest. The various Government bodies and private organisations noted in the previous paragraphs of this chapter may be contacted for information. In addition, details of training and courses are featured from time to time in *The Journal of Occupational Health and Safety — Australia and New Zealand* (see ¶1204).

A comprehensive listing of courses available (both short courses and tertiary courses) can be found in the *National Occupational Health and Safety Directory* (see ¶1204).

Tertiary courses in Occupational Health & Safety or Risk Management are available from many universities and TAFE colleges. The *course information office* at institutions in local areas can provide detailed information on course content and registration requirements. Many courses are part-time, lasting from two to four years.

Reference Section

Other CCH products in the area of OHS

Loose-leaf reporters

- *Plant Safety: Managing Plant Hazards in the Workplace*
- *Hazard alert: Managing Workplace Hazardous Substances*
- *Hazard alert: Hazardous Substances Induction Training Pack*
- *Laboratory Safety Manual*
- *Australian Workers Compensation Guide*
- *The Hands On Guide — OHS Legal Guide (includes CD-Rom)*
- *The Hands On Guide — OHS Manager (includes CD-Rom)*
- *The Hands On Guide — Risk Management (includes CD-Rom)*
- *Hands On Guide — School Health and Safety*
- *Safe Mining*
- *Victorian Accident Compensation Practice Guide*

CD-ROM

- *CCH Electronic Industrial Relations Law Library*
- *CCH Electronic OHS Library*

Training

- *Occupational Health and Safety Training Kit*

Books

- *Guidebook to Workers Compensation in Australia*
- *Occupational Health and Safety Laws in Australia*
- *Understanding New South Wales Occupational Health and Safety Legislation*
- *Workplace Rehabilitation Manual*
- *Lessons from Longford — The Esso Gas Plant Explosion*

¶1206

Index

References are to paragraph (¶) number.

A

Accident investigation, prevention and reporting 601
accidents
— causes 603
— common types 605
— definition 602
— prevention flowchart 606
— "proneness" and "blame" 604
disciplinary policy 618
employee feedback 617
investigation
— accident causes 610
— areas to be investigated 608
— role of supervisors 609
— special equipment 611
— steps involved 607
preventing recurrence of accidents 616
recording systems 612
reporting
— requirements 613
— writing skills 615
statistics 614

Accidents
complex causes 303
definition 109
ratio "pyramid" 103
statistics 102

ACT WorkCover 1202

Age and experience of workers
employer concerns 1118
issues 1117
risk management 1119

Alcohol and drug dependence
counselling 1110
detection of problem 1107
Employee Assistance Programs 1109
extent of problem 1106
role of employer 1108

Association of Risk and Insurance Managers of Australia (ARIMA) 1203

Audits
planning programs 405
risk elimination 805
scope 807
steps involved 806

Australasian Faculty of Occupational Medicine (AFOM) 1203

Australian and New Zealand Society of Occupational Medicine (ANZSOM) 1203

Australian Chamber of Commerce & Industry (ACCI) 1203

Australian Council of Trade Unions (ACTU) 1203

Australian Fire Protection Association 1203

INDEX

	Paragraph
Australian Institute of Occupational Hygienists	1203
Australian Nuclear Science and Technology Organisation (ANSTO)	1202
Australian Occupational Health Nurses Association (AOHNA)	1203
Australian Red Cross Society	1203

B

Back-up controls, definition	306

C

Checklists
fire prevention and control	931
manual handling risk factors	1002
occupational cancer	1007
planning programs	410

Codes
legislation	206
occupational hearing loss	1018

Comcare Australia	1202
Common law liability	207

Communicating health and safety programs
incentive schemes and competitions	713
occupational health service	714
posters	710
printed matter	711
promotion	709
videos and films	712

Communication
risk management	307

Company doctor
staffing	512

Consultative approach to health and safety
legislation	204
overview	106

Contributory negligence, definition	207

Costs
accidents	102; 103

D

	Paragraph
Danger tags	911

Definitions
accidents	109; 602
back-up controls	306
codes and standards	201
contributory negligence	207
"duty of care"	207
engineering controls	306
ergonomics	109
fire	922
hazard elimination	306
hazard substitution	306
hazards	302
health and safety policy	306
incidents	109
industrial awards and agreements	201
minimum standards compliance	101
occupational health	109
occupational rehabilitation	715
occupier's liability	207
OHS policy	306
regulations	201
risk	302
risk analysis	306
risk control	306
risk evaluation	306
risk identification	306
risk isolation	306
risk management	202; 301
statutes	201
vicarious liability	207

Department of Community Services and Health (Cth)	1202

Dermatitis
common causes	1011
diagnosis and treatment	1012

Disciplinary policy
accident investigation, prevention and reporting	618

"Duty of care", definition	207

E

Emergency planning
fire prevention and control	929
hazardous substances	916

Eme

Paragraph

Employees
- accident investigation, prevention and reporting feedback..........617
- evaluation of health and safety program..........804
- feedback..........618
- medical examinations..........720
- recruitment and selection in health and safety programs..........705
- role
  - health and safety..........106
  - risk management..........307
- training..........708

Employees with disabilities
- planning programs..........409

Employers
- role in programs
  - age and experience of workers........1118
  - alcohol and drug dependance........1108
  - alleviating stress..........1026

Enforcement
- legislation..........205

Engineering controls, definition..........306

Ergonomics
- definition..........109
- planning programs..........408

Ergonomics Society of Australia........1203

Ergonomists..........514
- staffing..........514

F

Fire prevention and control
- aspects..........921
- causes of fire..........923
- checklist..........931
- electronic equipment hazards..........925
- emergency evacuation plan..........929
- fire, definition..........922
- fire-drills..........930
- fire-fighting equipment..........926
- high-piled storage hazards..........924
- overall responsibility..........928
- warning systems..........927

First aid officers
- staffing..........509

Emp

Paragraph

Flowcharts
- accident prevention..........606

G

Group health service
- staffing..........515

H

Hazard elimination, definition..........306

Hazard substitition, definition..........306

Hazardous substances..........912
- control..........915
- emergency planning..........916
- fire control
  - electronic equipment..........925
  - high-piled storage..........924
- health effects..........913
- management..........914

Hazards
- types..........306

Hazards, definition..........302

Health and safety
- aims of program..........108
- changes in approach..........104
- definitions..........109
- fire prevention and control
  - overall responsibility..........928
  - warning systems..........927
- indirect costs..........103
- legislation
  - codes and standards..........206
  - common law liability..........207
  - consultation with employees..........204
  - enforcement..........205
  - modern approach..........202
  - summary..........203
  - types..........201
  - workers compensation..........208
- measuring of problem..........102
- officer/managers..........506
- policy, definition..........306
- reasons for integrated and planned approach..........101
- representatives..........507
- role
  - employees in consultative approach..........106

INDEX

Paragraph
- human resources/health and safety department.................. 107
- management...................... 105
- scope of book........................ 110
- staffing................................. 501
- company doctor................... 512
- ergonomists......................... 514
- first aid officer..................... 509
- group health service............ 515
- health and safety representatives..... 507
- health and safety workplace committees........................ 508
- need to use expert advice.... 510
- occupational health nurse.... 511
- occupational hygienist......... 513
- occupational physician........ 512
- organisational psychologists... 514
- rehabilitation counsellor..... 514
- responsibility................ 502–504
- risk managers...................... 514
- role of health and safety officer/manager................. 506
- role of human resources department....................... 505
- various approaches.............. 101
- workplace committees......... 508

Health and safety problems.................. 901
fire prevention and control
- aspects................................ 921
- causes of fire...................... 923
- checklist............................. 931
- electronic equipment hazards....... 925
- emergency evacuation plan......... 929
- fire, definition................... 922
- fire-drills.......................... 930
- fire-fighting equipment...... 926
- high-piled storage hazards.... 924
hazardous substances............ 912
- control............................. 915
- emergency planning........... 916
- health effects.................... 913
- management...................... 914
injuries from plant
- danger tags....................... 911
- lockout systems................. 911
- machine guarding.............. 909
- maintenance..................... 910

Paragraph
- purchasing........................ 908
- scope................................. 907
slips, trips and falls
- causes................................ 918
- management...................... 919
- scope................................. 917
- vehicle fleet safety............. 920
working environment............ 902
- lighting............................. 906
- temperature...................... 905
- ventilation........................ 904
- workplace layout............... 903

Health and safety programs
communication
- incentive schemes and competitions................... 713
- occupational health service.... 714
- posters.............................. 710
- printed matter.................. 711
- promotion......................... 709
- videos and films................ 712
evaluation of performance
- assessing results against original objectives....................... 803
- audits........................ 805–807
- choosing measurable objectives....... 802
- employees' right to know...... 804
- importance....................... 801
implementation.................... 701
- issuing of policy statement...... 702
- procedures, rules and work method statements..................... 704
- recruitment and selection of employees....................... 705
- sample policy statement.... 703
medical examinations........... 720
off-the-job............................ 719
planning
- audit................................. 405
- basic questions................. 402
- checklist........................... 410
- content............................. 406
- effect of job design........... 407
- employees with disabilities..... 409
- ergonomics....................... 408
- identification of specific organisational problems....................... 404

Hea

                                              Paragraph
Health and safety programs—continued
  – risk identification techniques.......... 401
  – role of employees............................ 403
  rehabilitation....................................... 716
  – definition... 715
  – features.. 717
  – State systems................................... 718
  training
  – continuation.................................... 708
  – induction... 707
  – purpose.. 706
Health and safety representatives
  legislation.. 507
  staffing.. 204
Health and safety workplace committees
  legislation.. 204
  staffing.. 508
Hierarchy of risk control...................... 306
Human resources department
  role in health and safety...................... 107
  staffing.. 505

                    I

Incidents, definition............................. 109
Indsutrial deafness — see Occupational hearing loss
Induction training................................ 707
Injuries from plant
  danger tags... 911
  lockout systems.................................. 911
  machine guarding............................... 909
  maintenance....................................... 910
  purchasing... 908
  scope.. 907

                    J

Job design
  planning programs.............................. 407
Job redesign
  manual handling injuries................. 1003

                    L

Legislation
  codes and standards........................... 206
  common law liability.......................... 207
  consultation with employees.............. 204
  enforcement....................................... 205

                                              Paragraph
  modern approach................................ 202
  summary.. 203
  types.. 201
  workers compensation........................ 208
Lighting... 906
Line managers
  staffing.. 504
Lockout systems.................................. 911

                    M

Machine guarding................................ 909
Management
  risk management................................ 307
  role in health and safety..................... 105
  training... 708
Medical examinations......................... 720
"Minimum standards compliance", definition.. 101

                    N

National Industrial Chemicals Notification and Assessment Scheme (NICNAS)........................ 912
National Occupational Health and Safety Commission................................... 1202
National Safety Council of Australia...................................... 1203
Networking
  risk management................................ 307

                    O

Obligations
  risk management................................ 302
Occupational back pain
  identification and assessment........... 1002
  prevention.. 1003
  regulation of manual handling.......... 1004
  scope.. 1001
Occupational cancer
  identification.................................... 1006
  prevention and control..................... 1007
  scope.. 1005
  skin cancer....................................... 1009
  synthetic mineral fibres.................... 1008
Occupational disease........................... 102
Occupational health, definition........... 109

Hea

Paragraph

Occupational health and safety — see **Health and safety**

Occupational health nurse
staffing.. 511

Occupational hearing loss
conservation programs..................... 1017
extent of problem............................... 1014
legislative provisions and
 standards....................................... 1018
measurement..................................... 1015
measurement of noise levels............ 1016

Occupational hygienist
staffing.. 513

Occupational overuse syndrome
nature.. 1019
reduction of causes........................... 1021
types of workers affected................. 1020

Occupational physician
staffing.. 512

Occupational physician/company doctor
staffing.. 512

Occupational skin diseases................ 1010
causes of dermatitis.......................... 1011
diagnosis and treatment of
 dermatiti.. 1012
preventive measures........................ 1013

Occupier's liability, definition............ 207

Off-the-job health and safety
 programs... 719

Operational planning............................ 105

Organisational psychologists
staffing.. 514

P

Physical health and safety problems
occupational back pain
 – identification and assessment......... 1002
 – prevention....................................... 1003
 – regulation of manual handling....... 1004
 – scope... 1001
occupational cancer
 – identification.................................. 1006
 – prevention and control.................. 1007
 – scope... 1005
 – skin cancer..................................... 1009

Paragraph

– synthetic mineral fibres................. 1008
occupational hearing loss
– conservation programs................... 1017
– extent of problem........................... 1014
– legislative provisions and
 standards.. 1018
– measurement................................... 1015
– measurement of noise levels.......... 1016
occupational overuse syndrome
– nature... 1019
– reduction of causes......................... 1021
– types of workers affected............... 1020
occupational skin diseases................. 1010
– causes of dermatitis........................ 1011
– diagnosis and treatment of
 dermatiti... 1012
– preventive measures....................... 1013
stress
– causes... 1023
– definition... 1022
– effects... 1024
– individual responses....................... 1025
– methods of alleviation.................... 1026

Planning
health and safety programs
– audit... 405
– basic questions................................ 402
– checklist... 410
– content... 406
– effect of job design......................... 407
– employees with disabilities............ 409
– ergonomics...................................... 408
– identification of specific organisational
 problems... 404
– risk identification techniques......... 401
– role of employees........................... 403
risk management................................. 305

Protective clothing and equipment
risk control.. 306

R

References... 1201
audio-visual aids................................ 1205
government organisations................ 1202
private organisations......................... 1203
publications....................................... 1204
training and tertiary courses............ 1206

	Paragraph
Regulations	
definition	201
manual handling	1004
occupational hearing loss	1018
Rehabilitation	
health and safety programs	716
– definition	715
– features	717
– State systems	718
legislation	208
Rehabilitation counsellor	
staffing	514
Risk analysis, definition	306
Risk control, definition	306
Risk evaluation, definition	306
Risk identification	
planning programs	401
Risk identification, definition	306
Risk isolation, definition	306
Risk management	
age of workers	1119
causes of accidents	303
definition	202; 301
elements	306
factors influencing success	307
planning	305
relationship with health and safety	304
"risks" and "hazards"	302
violence in the workplace	1114
Risk manager	
staffing	514
Risks, definition	302

S

	Paragraph
Safety Institute of Australia (SIA)	1203
Shift work and night work	
effects	1103
health and biological effects	1102
reducing ill effects	1105
stand-by periods	1104
types	1101
Slips, trips and falls	
causes	918
management	919

	Paragraph
scope	917
vehicle fleet safety	920
Smoking in the workplace	
passive smoking	1115
some solutions	1116
Social health and safety problems	
age and experience of workers	
– employer concerns	1118
– issues	1117
– risk management	1119
alcohol and drug dependence	
– counselling	1110
– detection of problem	1107
– Employee Assistance Programs	1109
– extent of problem	1106
– role of employer	1108
shift work and night work	
– effects	1103
– health and biological effects	1102
– reducing ill effects	1105
– stand-by periods	1104
– types	1101
smoking	
– passive smoking	1115
– some solutions	1116
violence	
– definition	1111
– incidence	1112
– likely targets	1113
– risk management	1114
St John Ambulance Association	1203
Staffing	501
ergonomists	514
first aid officer	509
group health service	515
health and safety representatives	507
health and safety workplace committees	508
need to use expert advice	510
occupational health nurse	511
occupational hygienist	513
occupational physician/company doctor	512
organisational psychologists	514
rehabilitation counsellor	514

INDEX

	Paragraph
responsibility	502
– line managers and supervisors	504
– top management	503
risk managers	514
role of health and safety officer/manager	506
role of human resources department	505

Standards
- legislation 206
- occupational hearing loss 1018

Standards Association of Australia 1203

Statistics
- accident investigation, prevention and reporting 614

Strategic planning 105

Stress
- causes 1023
- definition 1022
- effects 1024
- individual responses 1025
- methods of alleviation 1026

Supervisors
- accident investigation, prevention and reporting 609
- staffing 504

Surveys
- planning health and safety programs
  - checklist 410
  - employees with disabilities 409
  - role of employees 403

Synthetic mineral fibres 1008

T

Temperature 905

Trade Union Training Authority (TUTA) 1203

Training
- continuation 708
- induction 707
- manual handling injuries 1003
- outside courses 1206
- purpose 706
- stress 1026

U

Uniform health and safety legislation 202

V

Vehicle fleet safety 920
Ventilation 904
Vicarious liability, definition 207
Victorian WorkCover Authority 1202

Violence in the workplace
- definition 1111
- incidence 1112
- likely targets 1113
- risk management 1114

W

Work Health Authority (NT) 1202

WorkCover Authority of NSW 1202

WorkCover Corporation of South Australia 1202

Workers compensation 208

Working environment 902
- lighting 906
- temperature 905
- ventilation 904
- workplace layout 903

Workplace Health and Safety, Department of Training and Industrial Relations (Qld) 1202

Workplace layout 903

Workplace Standards Authority (Tas) 1202

WorkSafe Western Australia 1202